Soil Fertility and Fertilizers

Soil Fertility and Fertilizers

Virginia Munn

SYRAWOOD
PUBLISHING HOUSE
New York

Published by Syrawood Publishing House,
750 Third Avenue, 9th Floor,
New York, NY 10017, USA
www.syrawoodpublishinghouse.com

Soil Fertility and Fertilizers
Virginia Munn

International Standard Book Number: 978-1-64740-010-1 (Hardback)

Cataloging-in-Publication Data

Soil fertility and fertilizers / Virginia Munn.
p. cm.
Includes bibliographical references and index.
ISBN 978-1-64740-010-1
1. Soil fertility. 2. Fertilizers. 3. Soil productivity. 4. Soils--Fertilizer movement.
I. Munn, Virginia.
S596.7 .S65 2020
631.422--dc23

Printed in China.

TABLE OF CONTENTS

PREFACE

The ability of soil to sustain agricultural plant growth is referred to as soil fertility. A soil that is fertile has the ability to supply essential plant nutrients and water in appropriate amounts for the growth and reproduction of plants. It is also free of the toxic substances which could inhibit the growth of plants. A natural or synthetic material which supplies nutrients to plants which are essential for their growth is termed as a fertilizer. They generally provide macronutrients such as nitrogen, phosphorous and potassium in varying proportions. Fertilizers are broadly classified into single nutrient and multinutrient fertilizers. The topics included in this book on soil fertility and fertilizers are of utmost significance and bound to provide incredible insights to readers. Different approaches, evaluations, methodologies and advanced studies on this field have been included in it. This book will serve as a reference to a broad spectrum of readers.

Given below is the chapter wise description of the book:

Chapter 1- Soil is a mixture of minerals, gases, organic matter, liquids and organisms. Depending upon the spaces between its particles, soil is generally classified as permeable soil, water-holding soil or loam. This chapter has been carefully written to provide an easy understanding of the varied facets of soils.

Chapter 2- The ability of soil to sustain agricultural plant growth is known as soil fertility. It results in sustained and consistent high quality yields along with providing plant habitat. To maintain the soil fertility, it is very important to maintain the quality of irrigation water. This chapter discusses in detail the theories and methodologies related to soil fertility.

Chapter 3- Fertilizer is a material that is applied to soils or to plant tissues to supply one or more nutrients which are integral to the growth of the plant. Fertilizers can be natural or synthetic. The topics elaborated in this chapter will help in gaining a better perspective about the various types of fertilizers.

Chapter 4- Fertilizers play a major role in the enhancement and growth of plants. Nitrogen fertilizers, phosphorus fertilizers and potassium fertilizers are a few types of important fertilizers. The fertilizers made up of ammonia are known as nitrogen fertilizers. Phosphate fertilizers are obtained by extraction from minerals containing the phosphate ion. The topics elaborated in this chapter will help in gaining a better perspective about the different fertilizers and their manufacturing process.

Chapter 5- The science and practice of achieving optimal nutrient use efficiency, crop quality, economic returns, etc. while reducing off-site transport of nutrients is known as nutrient management. Soil health is a condition where a soil meets its range of ecosystem functions as appropriate to its environment. The chapter closely examines the key concepts of nutrition management and soil health to provide an extensive understanding of the subject.

At the end, I would like to thank all those who dedicated their time and efforts for the successful completion of this book. I also wish to convey my gratitude towards my friends and family who supported me at every step.

Virginia Munn

Chapter 1

Introduction to Soil

Soil is a mixture of minerals, gases, organic matter, liquids and organisms. Depending upon the spaces between its particles, soil is generally classified as permeable soil, water-holding soil or loam. This chapter has been carefully written to provide an easy understanding of the varied facets of soils.

Soils are complex mixtures of minerals, water, air, organic matter, and countless organisms that are the decaying remains of once-living things. It forms at the surface of land – it is the "skin of the earth." Soil is capable of supporting plant life and is vital to life on earth.

Soil, is:

1. The unconsolidated mineral or organic material on the immediate surface of the earth that serves as a natural medium for the growth of land plants.
2. The unconsolidated mineral or organic matter on the surface of the earth that has been subjected to and shows effects of genetic and environmental factors of climate (including water and temperature effects), and macro- and microorganisms, conditioned by relief, acting on parent material over a period of time.

Organisms in the soil need air and water to survive. Having these essential materials - air, water, and organic matter - makes it possible for plants, bacteria, fungi and small animals like earthworms and insects to live in the soil.

All the living things in the soil, plus essential materials that these organisms use to survive, form the soil ecosystem. Scientists study the soil ecosystem because they want to understand how organisms relate to one another and to the environment that surrounds them.

Dirt is what gets on our clothes or under our fingernails. It is soil that is out of place in our world – whether tracked inside by shoes or on our clothes. Dirt is also soil that has lost the characteristics that give it the ability to support life – it is "dead." Soil performs many critical functions in almost any ecosystem (whether a farm, forest, prairie, marsh, or suburban watershed). There are seven general roles that soils play:

1. Soils serve as media for growth of all kinds of plants.
2. Soils modify the atmosphere by emitting and absorbing gases (carbon dioxide, methane, water vapor, and the like) and dust.

3. Soils provide habitat for animals that live in the soil (such as groundhogs and mice) to organisms (such as bacteria and fungi), that account for most of the living things on Earth.

4. Soils absorb, hold, release, alter, and purify most of the water in terrestrial systems.

5. Soils process recycled nutrients, including carbon, so that living things can use them over and over again.

6. Soils serve as engineering media for construction of foundations, roadbeds, dams and buildings, and preserve or destroy artifacts of human endeavors.

7. Soils act as a living filter to clean water before it moves into an aquifer.

Soil Texture and Composition

The inorganic portion of soil is made of many different size particles, and these different size particles are present in different proportions.

A loam field

The combination of these two factors determines some of the properties of the soil:

1. A permeable soil allows water to flow through it easily because the spaces between the inorganic particles are large and well connected. Sandy or silty soils are considered 'light' soils because they are permeable, water-draining types of soils.

2. Soils that have lots of very small spaces are water-holding soils. For example, when clay is present in a soil, the soil is heavier, holds together more tightly, and holds water.

3. When a soil contains a mixture of grain sizes, the soil is called a loam.

When soil scientists want to precisely determine soil type, they measure the percentage of sand, silt, and clay. They plot this information on a triangular diagram, with each size particle at one corner. The soil type can then be determined from the location on the diagram. At the top, a soil would be clay; at the left corner, it would be sand, and at the right corner it would be silt. Soils in the lower middle with less than 50% clay are loams.

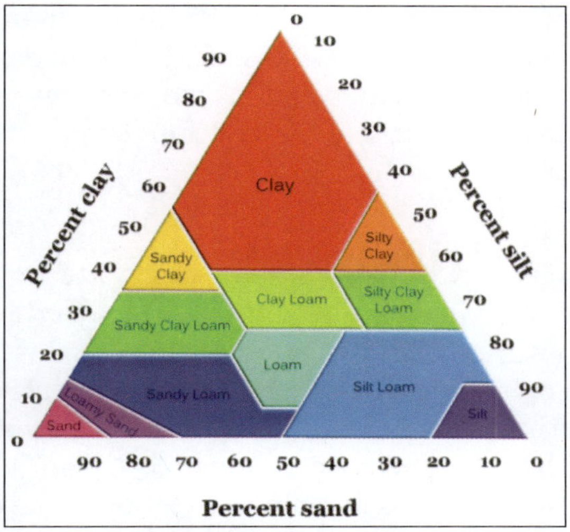

Soil types by particle size

Fundamentals of Soil

The soil that is cultivated to create a seedbed only consists to half of solid material, while the remainder consists of pores filled with water or air. Of the solid material, clay and organic material have the strongest influence on the soil and determine its tillage properties.

Soil consists of around 50% solid material, with 50% pores in between. In simple terms, half of a soil clod is solid material and the other half is pores.

Pores with Water or Air

The solid material consists of either mineral particles of different size classes or of organic material. The important pores are filled with either air or water depending on how wet the soil is at a particular time, its structure and soil tillage. In the ideal case, half the pores are filled with water and half with air. However, in soils with an aggregated structure such as clays, the pore volume is somewhat higher (40-60%) than in single-grained soils such as sands (35-45%).

The figure shows the difference in size between some of the mineral particles in a soil and the significance of this. Clay and humus particles are the smallest constituents in the soil. Their average diameter is less than 0.0002mm (i.e. 1000 times smaller than a grain of sand) and they are called colloids. The surface of clay particles has a negative electrical charge. This means that positively charged nutrients such as potassium, calcium and magnesium ions can bind to the clay particles. The clay particles therefore comprise nutrient reserves for plants.

Silt Holds Water

A cultivated arable soil is often a mixture of different particle size groups. If gravel and sand dominate the soil texture, this gives permeable, dry and relatively infertile soils, while an inclusion of sand in a clay soil makes it warmer. Silty soils are often cold and water-retaining and can easily take up water through capillary rise. The finest mineral particles, clay, have a strong influence on the soil even at concentrations of around only 5%. Clay soils shrink and swell and impart an aggregated stucture to the soil, with cracks and crevices where roots can grown through the soil profile. The typical characteristics of various soils are often a function of the clay content, which strongly influences soil type and tillage.

The Decomposition Process

1. Soil fauna start decomposition of dead organic material, partly through breaking it down into smaller pieces and partly through digging holes in the soil, increasing the supply of oxygen. Earthworms play a specific and very valuable role by breaking down the material and mixing it into the soil.
2. Bacteria and fungi continue decomposition in stages. The last phase, the formation of simple end-products that are available to plants (e.g. nitrate, phosphate and sulphate), is called mineralisation.
3. Humus formation. The decomposition of various organic compounds proceeds through a series of intermediate products that are of an increasingly simple nature as decomposition advances. These intermediate products react with each other and with the compounds created by soil organisms. This leads to the formation of new chemical compounds that are transformed into high molecular weight, dark-coloured matter known as humic substances. These humic substances have the ability to bind positively charged ions of potassium, calcium and magnesium, for example.

Smallest Particles have Large Specific Area

Fine clay (<0.0002mm) and some organic material are colloids and represent the smallest components in the soil. However, they have a large specific area, i.e. a large surface area in relation to their weight. The specific area increases with decreasing particle size. The surface of clay particles is negatively charged, so nutrients in the soil that are cations bind to the surface, creating a store of nutrients for the plants.

Particle Size Classes

The texture of the soil refers to the distribution of mineral particles into different size classes. Different countries use different classes, but a common internationally acceptable system classifies the texture into blocks, stones, gravel, sand, silt and clay according to the particle size ranges.

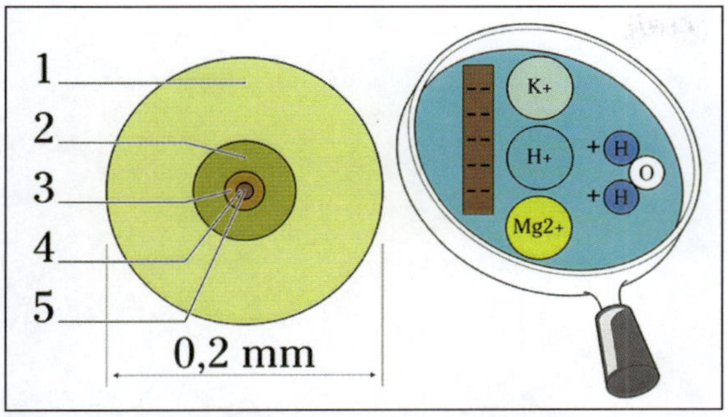

1) Fine sand 2) Very fine sand 3) Coarse silt 4) Fine silt 5) Coarse clay

Organic Material Positive

The organic material in the soil also has a very clear influence on the character of the soil. It consists to almost 60% of carbon (C) and originates from plant residues that have been decomposed by microorganisms. In this decomposition process, plant nutrients such as nitrogen (N), phosphorus (P) and sulphur (S) are released. The organic material can have enormous significance for soil properties and its influence is nearly always positive from the farmers perspective. It affects:

- Structure and aggregate stability.
- Water management.
- Tillage.
- Nutrient reserves.
- Silting and crusting.

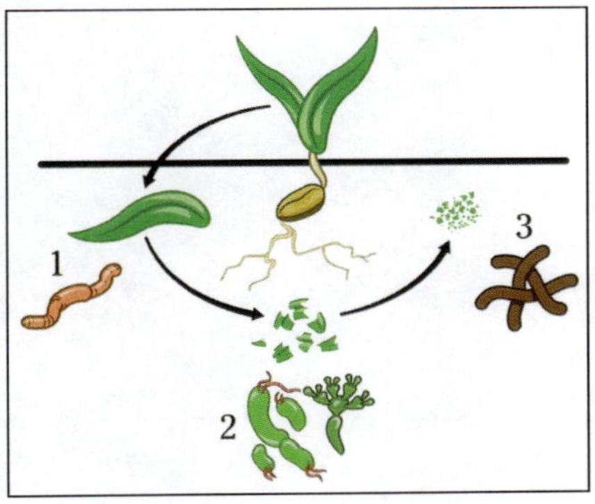

Particle Size Distribution Table

Particle group	Particle diameter (mm)
Clay	<0.002
Silt	0.002-0.06
Sand	0.06-2
Gravel	2-60
Stones	60-600
Blocks	<600

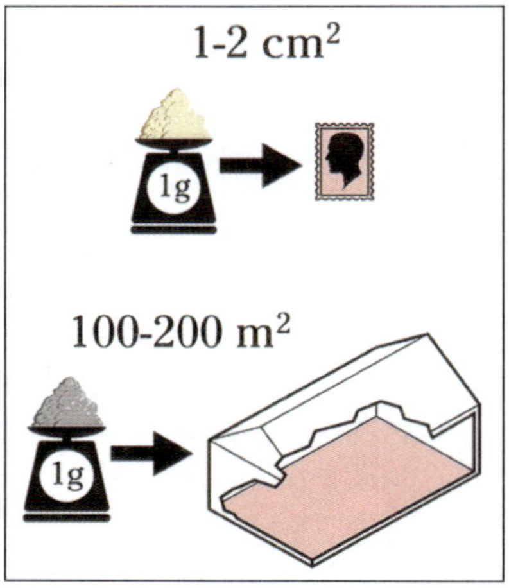

A characteristic of all clay minerals is their flattened shape. This, together with their extremely small size, means that clay colloids have a very large surface area in relation to their mass – a high specific area.

For example, one gram of sand has a combined area of around 1.5-2 cm^2, which is equivalent to a small stamp. However, one gram of clay can have a combined area of several 100 m^2 – an average-sized house.

Soil Structure

Soil structure describes the physical configuration of the soil. The sand particles in a sand soil are held together weakly and do not form aggregates, while the clay particles in a clay soil readily form aggregates. These aggregates make clay soil easy to till and improve the transport of air and water.

Soil structures are defined as the 3-D arrangement of primary particles in the soil, i.e. how the primary particles are situated and connected to each other. Determining factors for soil structure include its texture and content of organic material. A distinction is made between soils with a single-grained structure and soils with an aggregated structure.

Sand Soils are not Held Together

A sand soil such as this is an example of a single-grained structure. The grains of sand are relatively large and often only held together weakly. Even sand soils with a higher content of colloids fall apart easily when exposed to pressure in the soil. A sand soil with a low clay content often requires deeper cultivation in order to create a good arable soil and growing site. The low clay content gives the sand soil a low internal structure forming capacity.

A sand soil is an example of a single-grained structure

A clay soil such as this is held together and often has an aggregated structure

Clay Soils becomes Aggregated

A clay soil such as this is held together and often has an aggregated structure. Even at a content of 5%, the clay has a very strong impact on the soil and dominates its properties.

The aggregated structure is the result of a range of processes in the soil that in combination form aggregates. These structure forming processes affect the soil structure in a dynamic interplay with soil tillage.

When clay particles are bound together in aggregates, the properties of the soil are almost always improved

Aggregates Improve the Soil

When clay particles are bound together in aggregates, the properties of the soil are almost always improved. An important effect is that soil tillage becomes easier. The movement of air is also better, improving the transport of oxygen to and carbon dioxide from the root system. The aggregate structure also improves soil permeability to water and its water-holding capacity. Good soil structure makes an arable clay soil a more productive growing site, since all essential functions for the plant and its root system are provided. This gives faster growth and higher yield.

Soil Structure Forming Processes

The structure of a clay soil is the result of a number of different processes, which together give the soil profile its character. Soil tillage also affects soil structure through disturbing it to some depth and mixing in harvest residues.

Frost or Heat

Frost and cold in the winter and heat in the summer have the same effect on the soil – removal of water. Both processes dry out the soil and when water is removed the clay particles are pressed closer to each other. The result is mechanically created aggregates.

Chemical Processes

Organic material, iron and aluminium oxides and carbonates stabilise the aggregates through acting as binding agents. In soils that are not tilled, this process plays a major role. However, repeated soil tillage decreases the stability of the aggregates and in an arable soil the structure is therefore more dependent on soil clay content and biological activity.

Earthworms

Earthworms eat plant remains and mix them into the soil as they munch their way through the soil profile. Their casts act as adhesive between the soil particles and increase aggregate stability.

Earthworms also stimulate the soil microorganisms, which increase soil stability through producing mucus and other binding compounds.

Plants

Plants dry out the soil through their water uptake in the same way as frost and heat. When the soil dries out, the clay particles are forced closer to each other, creating and strengthening aggregates. Plant roots also increase the amount of organic material in the soil and leave root channels behind. Overall, this means that a plant cover on soil builds up the soil structure, while a bare soil without growing plants breaks it down.

Drainage

Drainage transports away excess water in the soil profile and thus dries out the soil. In this way, drainage helps to improve the soil structure. Very few soils are naturally self-draining and therefore drainage is essential for uniform drying of the soil, decreased soil compaction and early establishment of crops in autumn and spring.

Organic Material and Lime

Animal manure, intercrops, harvest residues etc. supply organic material to the soil. In the short term, this increases biological activity and causes a bloom of soil microorganisms. In the long term, it increases the content of organic material, which improves aggregate stability. Regular addition of lime also improves aggregate formation.

Heavy Machinery

On arable soils, passes with heavy machinery often cause soil compaction. Soil compaction results in the large pores in the soil being compressed and eliminated. This means that rapid transport of water and air to the roots is impeded. Drainage is also restricted and it becomes more difficult for the roots to force their way down through the soil.

The Outcome is the Soil Profile

Together, all the structure forming processes described above result in a soil profile that often has finer aggregates at the surface and coarser aggregates further down. The processes are dynamic and therefore the structure varies over time. For farmers with their sights set on the future and sustainable production, it is important to understand the significance of factors controlling the structuring process. Through diagnosis, it is rather easy to get a picture of soil structure.

Tillage affects Structure

Cropping system affects the soil structure and the placement of plant residues determines whether to employ conventional or minimal tillage or even direct drilling. The tilled layer often has a looser structure, while a denser layer is created at tillage depth. This denser layer develops regardless of tillage technique used, but the depth at which it arises varies. To avoid creating dense layers in the soil, it is important to vary tillage depth.

Types of Soil

On the basis of Climate Type

Although soil scientists recognize thousands of types of soil—each with its own specific characteristics and name—let's consider just three soil types. This will help you to understand some of the basic ideas about how climate produces a certain type of soil.

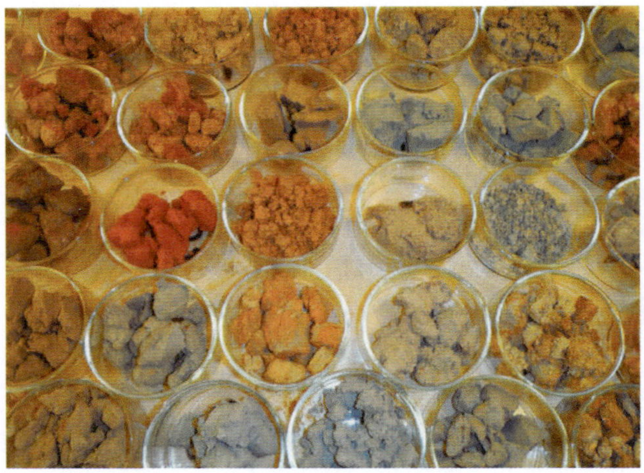

Just some of the thousands of soil types

Pedalfer

Deciduous trees, the trees that lose their leaves each winter, need at least 65 cm of rain per year. These forests produce soils called pedalfers, which are common in many areas of the temperate, eastern part of the United States. The word pedalfer comes from some of the elements that are commonly found in the soil. The Al in pedalfer is the chemical symbol of the element aluminum, and the Fe in pedalfer is the chemical symbol for iron. Pedalfers are usually a very fertile, dark brown or black soil. Not surprising, they are rich in aluminum clays and iron oxides. Because a great deal of rainfall is common in this climate, most of the soluble minerals dissolve and are carried away, leaving the less soluble clays and iron oxides behind.

A pedalfer is the dark, fertile type of soil that will form in a forested region

Pedocal

Pedocal soils form in drier, temperate areas where grasslands and brush are the usual types of vegetation. The climates that form pedocals have less than 65 cm rainfall per year, so compared to pedalfers, there is less chemical weathering and less water to dissolve away soluble minerals so more soluble minerals are present and fewer clay minerals are produced. It is a drier region with less vegetation, so the soils have lower amounts of organic material and are less fertile.

A pedocal is the alkaline type of soil that forms in grassland regions

A pedocal is named for the calcite enriched layer that forms. Water begins to move down through the soil layers, but before it gets very far, it begins to evaporate. Soluble minerals, like calcium carbonate, concentrate in a layer that marks the lowest place that water was able to reach. This layer is called caliche.

Laterite

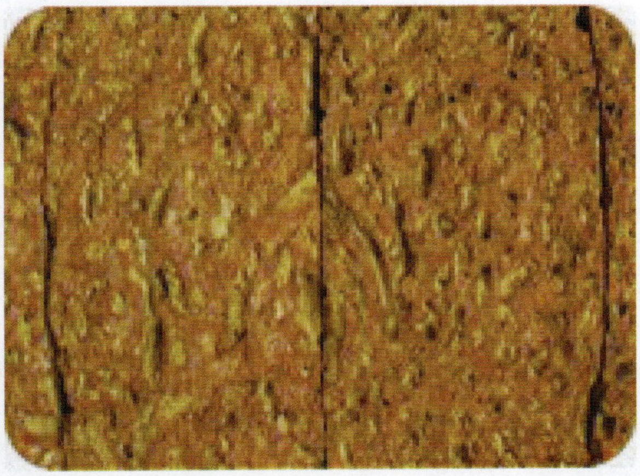

A laterite is the type of thick, nutrient poor soil that forms in the rainforest

In tropical rainforests where it rains literally every day, laterite soils form. In these hot, wet, tropical regions, intense chemical weathering strips the soils of their nutrients. There is practically no humus. All soluble minerals are removed from the soil and all plant nutrients are carried away. All that is left behind are the least soluble materials, like aluminum and iron oxides. These soils are often red in color from the iron oxides. Laterite soils bake as hard as a brick if they are exposed to the sun.

Many climates types have not been mentioned here. Each produces a distinctive soil type that forms in the particular circumstances found there. Where there is less weathering, soils are thinner but soluble minerals may be present. Where there is intense weathering, soils may be thick but nutrient poor. Soil development takes a very long time, it may take hundreds or even thousands of years for a good fertile topsoil to form. Soil scientists estimate that in the very best soil-forming conditions, soil forms at a rate of about 1mm/year. In poor conditions, soil formation may take thousands of years.

Characteristics of Different Soil Types

The soil types determine how and what we can grow and are the basis for all farming. Here is a quick guide to the characteristics of each soil type.

Sand Soils

Sand soils are often dry, nutrient deficient and fast-draining. They have little (or no) ability to transport water from deeper layers through capillary transport. Therefore; tillage of sandy soils in the spring should be kept to a minimum in order to retain moisture in the seedbed. The nutrient- and water-holding capacity of sand soils can be improved through adding organic material.

Silt Soils, 0-10% Clay

These soils differ from sand soils by having a greater tendency to form a crust, which is often very hard. If they are over-tilled, they can become compact and this decreases their ability to infiltrate water in wet periods. In dry conditions they can become hard and difficult to till. However, they are generally easy to till and can store considerable amounts of water. They require good reconsolidation, but tillage in wet conditions should be avoided.

Clay Soils with 10-25% Clay

These soils differ from those described above in that crusting can be very severe. The crust is often so hard that it has to be broken up. With low contents of clay and organic material, aggregate formation is often poor.

Clay Soils with 25-40% Clay

These soils have a good ability to transport water by capillary action from deep layers but the rate is slow, so plant water requirements are not met through capillary water. These soils are darker in colour and soil aggregation is more distinct. Aggregation decreases the risk of crusting. These soils must be tilled at the correct water content in order to be easily cultivated. There is a risk of clodding if conditions are too dry, or of smearing if they are too wet. These soils have a good ability to improve their structure through the action of climate, roots etc.

Clay Soils with 40% Clay

Heavy clays have a very high water-holding capacity, but most of the water is tightly bound and not available to plants. The humus content is often higher than in other mineral soils. They do not form a crust when they dry. These soils have a very good ability to improve their structure through e.g. freezing/thawing and drying/wetting. In cold winters the clay freezes apart and forms a very favourable aggregated structure in the topsoil layer. If the clay dries out without having been frozen, it can become very stiff and difficult to work.

In the water-saturated state these soils can be sticky and very impermeable to water. Due to the high clay content, the nutrient content is very high. Heavy clays need a high degree of recompaction around the seed when they are dry, but not when they are damp and plastic. The risk with tilling them in wet conditions is that it leads to soil compaction.

Soil Acidity

An acid is a substance that tends to give up protons (hydrogen ions) to some other substance. Conversely, a base in any substance that tends to accept protons (hydrogen ions). Soil acidity may be defined as the soil system's proton (H^+ ions) donating capacity during its transition from a given state to a reference state.

Soil acidity involves intensity and quantity aspects. The intensity aspect is universally characterised by the measurements of H^+ ion activity, expressed as pH. The quantity aspect is characterised, directly or indirectly, by the quantity of alkali required to titrate soil to some arbitrarily established endpoint. Soil acidity is a major problem in relation to plant growth and therefore, acid soils are called a problem soil.

Source of Soil Acidity

Acid soil is a base unsaturated soil which has got enough of adsorbed exchangeable H^+ ions so that to give soil a pH of lower than 7.0. The following important sources which are responsible for the development of acidic soils.

(i) Leaching due to Heavy Rainfall:

Generally acid soils are common in all regions where rainfall or precipitation is high enough to leach appreciable amounts of exchangeable bases from the surface soils and relatively insoluble compounds of Al and Fe remains in soil. The nature of these compounds are acidic and its oxides and hydroxides react with water (H_2o) and release hydrogen (H^+)ions in soil solution and soil becomes acidic.

Besides, when the soluble bases are lost, the H^+ ions of the carbonic acid and other acids developed in the soil replace the basic cations of the colloidal complex. As the soil gets gradually depletes of its exchangeable bases through constant leaching, it gets de-saturated and becomes increasingly acid.

$$CO_2+H_2O=H_2CO_3$$

$$H_2CO_3+C_aCO_3=\underset{\text{leachable}}{Ca(HCO_3)_2}\downarrow$$

(ii) Acidic Parent Material:

Some soils have developed from parent materials which are acid, such as granite and that may contribute to some extent soil acidity.

(iii) Acid Forming Fertilizers and Soluble Salts:

The use of ammonium sulphate $(NH_4)_2SO_4$ and ammonium nitrate, NH_4NO_3 increases soil acidity. Ammonium (NH_4) ions from when applied to the soil replace calcium (Ca^{2+}) ions from the exchange complex and the calcium sulphate ($CaSO_4$) is formed and finally leached out.

$$(NH_4)_2\,SO_4 \rightleftharpoons 2NH_4^+ + SO_4^{2-}$$

$$2NH_4+Ca\boxed{Clay} \rightarrow \underset{\text{Leached Out}}{CaSO_4\downarrow} + \begin{matrix}NH_4\\NH_4\end{matrix}\boxed{Clay}$$

$$\begin{matrix}NH_4\\NH_4\end{matrix}\boxed{Clay} + 3O_2 \xrightarrow{\text{Nitrification}} \underset{\text{Acid soil}}{\begin{matrix}H\\H\end{matrix}\boxed{Clay}} + 2HNO_3$$

Besides, basic portion of ammonium sulphate, $(NH_4)_2SO_4$ is NH4 and it undergoes biological transformation in the soil and form acid forming nitrate (NO_3^-) ions. Similarly, sulphur also produce acid forming sulphate (SO_4^{2-})ions through oxidation.

Divalent cations of soluble salts usually have a greater effect on lowering soil pH than monovalent metal cations.

(iv) Humus and Other Organic Acids:

Humus materials in soils occur as a result of microbiological decomposition of organic matter and contain different functional groups like carboxylic (-COOH) phenolic (-OH) etc. which are capable of attracting and dissociating hydrogen (H^+)ions.

During organic matter decomposition, humus, organic acids and different acid salts may also be produced and also concentration of CO_2 increased. The increased concentration of CO_2 hydrolysis of acid salts and various organic acids combinedly increased the total acidity of soil.

(v) Aluminosilicate Minerals:

At low pH values most of the aluminium (Al) is present as the hydrated aluminium ions (Al^{3+}) which undergoes hydrolysis and release hydrogen (H^+) ions in the soil solution.

$$Al^{3+} + H_2O \overset{\text{Hydrolysis}}{\rightleftarrows} Al(OH)^{2+} + H^+$$
$$Al(OH)^{2+} + H_2O \rightleftarrows Al(OH)_2^+ H^+$$

It is also possible that structural OH^- ions at corners and edges may dissociate hydrogen (H^+) ions and develop soil acidity.

(vi) Carbon Dioxide (CO_2):

Soil containing high concentration of CO_2, the pH value of such soil will be low i.e. the soil becomes acidic. Root activity and metabolism may also serve as sources of CO_2 which ultimately helps the soil to become acidic.

(vii) Hydrous Oxides:

These are mainly oxides of iron and aluminium. Under favourable conditions they undergo step-wise hydrolysis with the release of hydrogen (H^+) ions in the soil solution and develop soil acidity.

(viii) Aluminium and Iron Polymers:

The $(Al)^{3+}$ ions displaced from clay minerals by cations are hydrolyzed to monomelic and polymeric hydroxy aluminium complexes. Hydrolysis of the monomelichexaquoaluminium or iron forms are illustrated by the following stepwise reactions with the liberation of hydronium ion (H_3O^+) and lower soil pH. Each successive step occurs at a higher pH:

$$Al(H_2O)_6^{3+} + \rightleftarrows Al(OH)(H_2O)_5^{2+} + H_3O^+$$
$$Al(OH)(H_2O)_5^{2+} + H_2O \rightleftarrows Al(OH)_2(H_2O)_4^+ + H_3O^+$$
$$Al(OH)_2(H_2O)_4^+ + H_2O \rightleftarrows Al(OH)_3(H_2O)_3^o + H_3O^+$$
$$Al(OH)_3(H_2O)_3^o + H_2O \rightleftarrows Al(OH)_4(H_2O)_2^- + H_3O^+$$

Similarly,

$$Fe(H_2O)_6^{3+} + H_2O \rightleftarrows Fe(OH)(H_2O)_5^{2+} + H_3O^+$$

(reaction occurs under more acidic condition than Al)

$$Fe(OH)_3(H2O)_3^o + H_2O \rightleftarrows Fe(OH)_4(H2O)_2^- + H2O^+$$

Kinds of Soil Acidity

Broadly soil acidity may be of two kind's viz:

(i) Active acidity and

(ii) Potential reserve/exchange acidity.

The nature of soil can be illustrated as follows:

$$\underset{\text{(potential/exchange/acidity)}}{\text{Adsorbed H(and Al) icons on soil colloids}} \rightleftarrows \underset{\text{(active acidit)}}{\text{Soil solution H(and Al) ions}}$$

Active Acidity

Active acidity may be defined as the acidity develops due to hydrogen (H^+) and aluminium $(Al)^{3+}$ ions concentration of the soil solution. The magnitude of this acidity is limited.

Exchange Acidity

Exchange acidity may be defined as the acidity develops due to adsorbed hydrogen (H^+) and aluminium $(Al)^{3+}$ ions on the soil colloids. The magnitude of this exchange acidity is very high. However, residual acidity may be included to the total acidity.

Residual acidity may be defined as the acidity which remains in the soil after active and exchange acidity has been neutralized. It is associated with aluminium hydroxy ions and with H and Al atoms that are bound in non-exchangeable forms by organic matter and silicate clay.

An equilibrium relationship between exchange and active acidity on 2:1 type colloid is depicted:

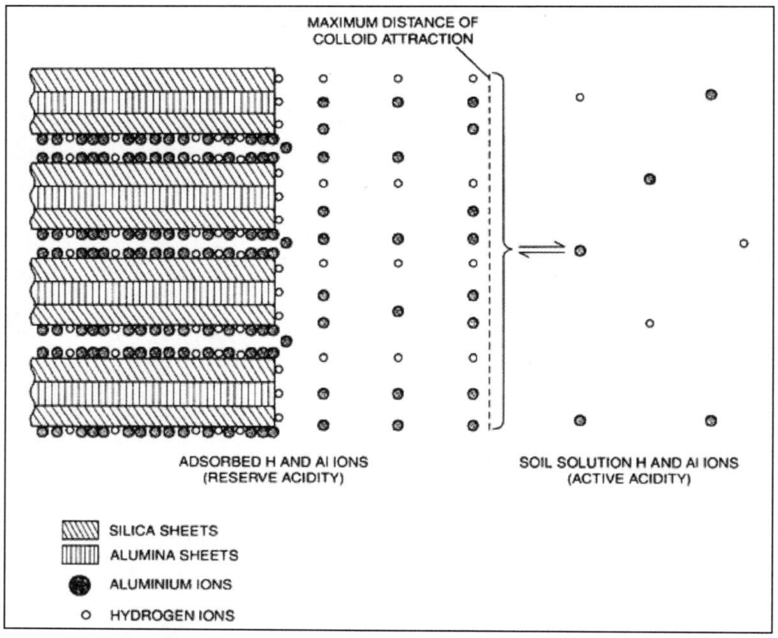

Relationship between reserve or exchange and active acidity

Total Acidity

The total acidity is summation of active, exchange and residual acidity. It can be written as,

Total acidity = Active acidity + Exchange acidity + Residual acidity.

Therefore, total soil acidity depends on the active, exchange and residual acidity of the soil.

Active acidity – Soluble acidity, in the solution.

Reserve acidity – Adsorbed acidity, on the surface of particles.

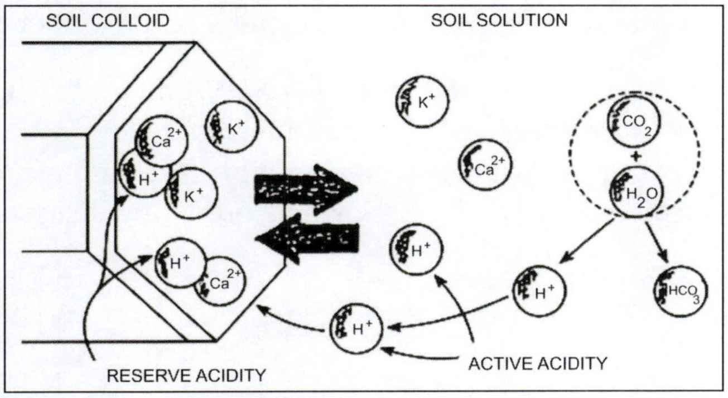

Adsorbed acidic ions: H^+, Al^{3+}, Fe^{3+}

Why are the trivalent ions acidic?

$$[Al(H_2O)_6]^{3+} = [Al(OH)(H_2O)_5]^{2+} + H^+$$

$$[Al(OH)(H_2O)_5]^{2+} = [Al(OH)_2(H_2O)_4]^{+} + H^+$$

$$[Al(OH)_2(H_2O)_4]^{+} = [Al(OH)_3(H_2O)_3] + H^+$$

$$[Al(OH)_3(H_2O)_3] = [Al(OH)_4(H_2O)_2]^{-} + H^+$$

$$[Al(OH)_4(H_2O)_2]^{-} = [Al(OH)_5(H_2O)]^{2-} + H^+$$

$$[Al(OH)_5(H_2O)]^{2-} = [Al(OH)_6]^{3-} + H^+$$

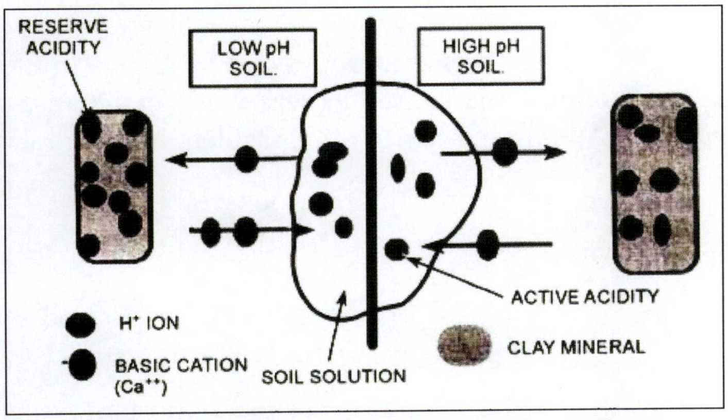

Diagram of soil pH and active and reserve acidity

Problems of Soil Acidity

Problems of soil acidity may be divided into three groups which are as follows:

Toxic Effects

1. Acid Toxicity: The higher hydrogen ion concentration is toxic to plants under strong acid conditions of soil. The acid toxicity includes possible toxicities of acid anions as well as H^+ ions.

2. Toxicity of different nutrient elements: Iron and Manganese - The concentration of these two ions (Fe^{2+} and Mn^{2+}) in soil solution depends upon the soil reaction or pH, organic matter and intensity of soil reduction. Due to increase in organic matter content in the soil the population of soil microbes increases and very rapidly used up the soil oxygen and results reduction of soil.

As a result of soil reduction, the nutrient elements like Mn^{4+} and Fe^{3+} reduce to Mn^{2+} (manganous manganese) and Fe^{2+} (ferrous iron) respectively and increases their concentration to a very high and toxicity of those elements develops. Due to such toxic effects, a physiological disease of rice is found in submerged soils which are popularly known as browning disease.

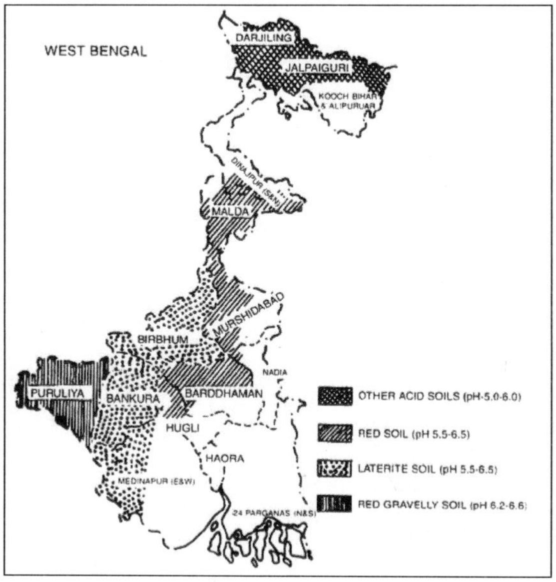

Acid soil map of west Bengal

Toxicity of Aluminium (Al) - The toxicity of aluminium may be greatly influenced by the accompanying cations. The toxicity of aluminium tends to decrease with an increase in the concentration of other cations such as calcium. Aluminium toxicity is a problem in both upland and lowland soils.

Aluminium toxicity in soils affects plant growth in various ways:

1. It restricts the root growth.
2. It affects various plant physiological processes like division of cells, formation of DNA and respiration etc.
3. It restricts the absorption and translocation of some important nutrient elements from soil to the plant like phosphorus, calcium, iron, manganese etc.
4. It causes wilting of plants.
5. It also inhibits the microbial activity in the soil.

Nutrient Availability

1. Non-specific Effects: It is associated with the inhibition effect of root growth and thereby affects the nutrient availability.

2. Specific Effects:

- Exchangeable Bases: There are two aspects of availability of exchangeable bases i.e. ion uptake process and the release of bases from the exchangeable form may be adversely affected due to soil acidity. Due to complementary ion effect exchangeable bases are released preferentially in a fractional exchange. Deficiency of bases like Ca^{2+} and Mg^{2+} are found in acid soils.
- Nutrient Imbalances: It is evident that soluble iron, aluminium and manganese are usually present in their higher concentrations under moderate to strong acid soils. Phosphorus reacts with these ions and produces insoluble phosphatic compounds rendering phosphorus unavailable to plants. Besides these, fixation of phosphorus by hydrous oxides of iron and aluminium or by adsorption, the availability of phosphorus is decreased.

In acid soils, iron, manganese, copper and zinc are abundant, but molybdenum is very limited and unavailable to plants. In acid soils having very low pH, the availability of boron may also be decreased due to adsorption on sesquioxides, iron and aluminium hydroxy compounds. Nitrogen, potassium and sulphur become less available in an acid soil having pH less than 5.5.

Microbial Activity

It is well-known that soil organisms are influenced by fluctuations in the soil reaction. Bacteria and actinomycetes function better in soils having moderate to high pH values. They cannot show their activity when the soil pH drops below 5.5. Nitrogen fixation in acid soils is greatly affected by lowering the activity of Azotobacter sp.

Besides these, soil acidity also inhibits the symbiotic nitrogen fixation by affecting the activity of Rhizobium sp. Fungi can grow well under very acid soils and caused various diseases like root rot of tobacco, blights of potato etc.

Amelioration of Soil Acidity

In general the fertility status of acid soils is very poor and under strongly to moderately acidic soils the plant growth and development affect to a great extent. The crops grown on such problematic soils do not give remunerative return rather it lowers down the yield to a great extent.

Because of the limited land resource it needs judicious management practices so that yield of different crops can be increased. So one of the most important and practically feasible management practices is the use of lime and liming materials to ameliorate the soil acidity.

The addition of lime raises the soil pH to some prescribed value. This value is usually in the range of pH to some prescribed value. This value is usually in the range of pH 6.0 to 7.0, since this is an easily attainable value within the optimum range of most crop plants.

Lime requirement of an acid soil may be defined as the amount of liming material that must be added to raise the pH to some prescribed value. This value is usually in the range of pH 6.0 to 7.0, since this is an easily attainable value within the optimum range of most crop plants.

Liming Reactions

Lime reactions in soils depend upon the nature and the fineness of the liming materials. Lime is usually applied to soils in the form of ground limestone. Limestone's can be classified as calcitic ($CaCO_3$) dolomite [$CaMg(CO_3)_2$] or a mixture of the two.

Both of these limestone's are sparingly soluble in pure water but do become soluble in water containing carbon dioxide. The greater the partial pressure of carbon dioxide in the system, the more soluble the limestone becomes. Dolomite is somewhat less soluble than calcite.

$$CaCO_3 + H_2O + CO_2 \rightarrow Ca(HCO_3)_2$$

$$Ca(HCO_3)_2 \rightarrow \underset{\text{(takes part in cation exchange reaction)}}{Ca^{2+}\downarrow} + 2HCO_3^-$$

$$\underset{\text{(from soil solution)}}{H^+} + \underset{\text{(from line)}}{HCO_3^-} \rightarrow H_2CO_3 \rightleftarrows H_2O + CO_2$$

In this way hydrogen ions (H^+) in the soil solution react to form weakly dissociated water, and the calcium (Ca^{2+}) ion from limestone is left to undergo cation exchange reactions. The acidity of the soil is, therefore, neutralized and the per cent base saturation of the colloidal material is increased.

The process of changing pH by the addition of lime [$Ca(OH)_2$] is illustrated in figure below:

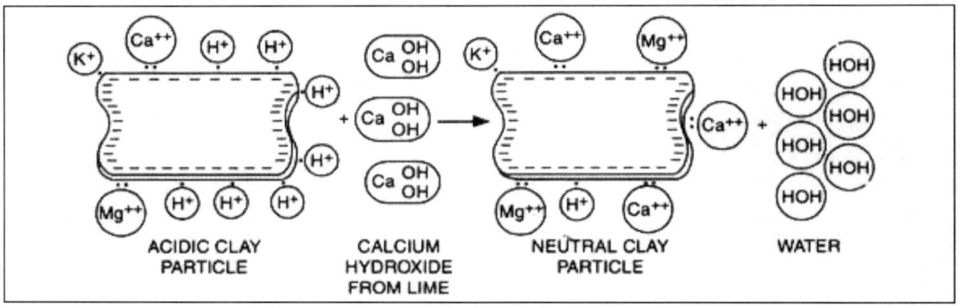

Change of pH with the application of time

Factors affecting Liming Reactions

Various environmental factors affect the rate of limestone reaction.

A few of the important factors are being discussed here:

- Moisture

The greater the amount of moisture, the more rapid is the rate of reaction. Obviously moisture must be present before the solubility reaction can occur. As moisture increases, the degree of aeration are reduced resulting an increase in the concentration of carbon dioxide (CO_2) in the soil air and thereby increases the rate of reaction. The increased moisture would also allow for a greater volume of solution and, therefore, a lower concentration of reaction end products. Since the reaction is an equilibrium reaction, the accumulation of end products would reduce reaction rate over time:

- Temperature: Lime and liming materials react more rapidly at high that at low temperatures. This effect is probably related to diffusion rates of end products away from the reaction sites.

- Amount of exchange acidity: The amount of exchange acidity present in the soil affects reaction rate. If a soil has a high lime requirement and if a sufficient quantity of limestone is added to neutralize the acidity present, the initial reaction will be quite rapid. However, as the acidity becomes neutralized, the rate of reaction decreases and finally, as neutrality is approached becomes almost negligible.

Liming Materials

Liming materials may be defined as materials that are necessary for the neutralization of soil solution hydrogen (H^+) ions. The materials commonly used for the liming of soils are the oxides, hydroxides, carbonates and silicates of calcium or calcium and magnesium.

The presence of only these elements does not consider a material as a liming compound. In addition to these compounds, the accompanying anion must be one that will reduce the activity of hydrogen (H^+)ions and hence aluminium in the soil solution. These are called "Agricultural liming materials".

Why Gypsum is not Considered as a Liming Material?

Gypsum is not considered as liming materials because on its application to an acid soil it dissociates into calcium (Ca^{2+}) and sulphate (SO_4^{2-}) ions:

$$CaSO_4 D Ca^{2+} + SO_4^{2-}$$

The accompanying anion is sulphate and it reacts with soil moisture produces mineral acid (H_2SO_4) which also increases soil acidity instead of reducing soil acidity.

$$SO^{4^{2-}} + H_2O \text{ ® } \underset{\text{(strong mineral acid)}}{H_2SO_4}$$

Besides this, calcium (Ca^{2+}) in the gypsum after dissociation will result in replacement of adsorbed aluminium (Al^{3+}) in a localized soil zone (when gypsum applied as band placement) with a significant lowering of soil pH. Therefore, gypsum does not qualify as a liming material.

Kinds of Liming Materials

There are various kinds of liming materials that are used for the correction of soil acidity.

Oxides of Lime

It is normally called burned lime or quick lime. Oxide of lime is more caustic than limestones.

Burned lime is produced by heating limestone and dolomite as follows:

$$\underset{\text{(Limestone)}}{CsCO_3} + Heat \rightarrow CaO + CO_2 - \uparrow$$

$$CaMg(CO_3)heat \rightarrow CaO + MgO + 2CO_2 - \uparrow$$

Hydroxides of Lime

It can be produced by adding water to burned lime and is called slaked lime.

$$CaO + H_2O \rightarrow Ca(OH)_2$$

It is more caustic than burned lime (CaO).

If it is kept open in the moist air, then combination of calcium hydroxide occurs as follows:

$Ca(OH)_2 + Co_2$â†' $CaCO_3 + H_2O$

In case of $Mg(OH)_2$

$Mg(OH)_2 + CO_2$â†' $MgCO_3 + H_2O$

Carbonates of Lime

These are by products of certain industries and so the content of calcium and magnesium varies. The two important minerals are found in this group- Calcite ($CaCO_3$) and dolomite [$CaMg(CO_3)_2$].

Slags

These are generally three types of slags that are found important:

Blast Furnace Slag

It is a by-product of the manufacture of pig iron. As a liming material, this slag behaves essentially as calcium silicate. The neutralizing value of blast furnace slags ranges from about 75-90%.

Basic Slag

It is a by-product of the basic open-hearth method of making steel from pig iron, which in turn is produced from high phosphorus iron ores. The impurities in the iron, including silica and phosphorus are fluxed with lime and the basic slags are produced. Its neutralizing value ranges from 60-70%.

Electric Furnace Slag

This is produced from the electric furnace reduction of phosphate rock during preparation of elemental phosphorus. This product is largely calcium silicate and is used as a liming material.

Other Liming Materials

Coral shell, chalk, wood ash, press mud, by-product material of paper mills, sugar factories, fly ash and sludge etc. are considered as liming materials and also used for the amelioration of soil acidity.

Efficiency of Liming Materials

The efficiency of liming materials can be judged on the basis of following important factors because of varying neutralizing capacity of different liming materials:

(1) Neutralizing value (N.V.) or calcium carbonate equivalent of liming materials,

(2) Purity of liming materials and

(3) Degree of fineness of liming materials.

(1) Neutralizing Value (N.V.) or Calcium Carbonate Equivalent (CCE): Calcium carbonate equivalent (CCE) is defined as the *acidâ€”*"neutralizing capacity of an agricultural liming material expressed as a weight percentage of calcium carbonate.

$$\text{CCE of limiting material} = \frac{\text{Molecular weight of CaCO}_3}{\text{Molecular weight of a limiting material whose CCE is to be determined}} \times 100$$

Example: Calculate the CCE of dolomite, $CaMg(CO_3)_2$.

Solution: In dolomite, there is mixture of $CaCO_3$ and $MgCO_3$. Both are almost equally effective in neutralizing soil acidity. Considering the materials as two separate liming materials present in the dolomite. So the molecular weight of dolomite may be considered as 184/2 = 92 since 84 gms of $MgCO_3$ will neutralize the same amount of acid as 100 gms of $CaCO_3$.

So,

$$\text{CCE of dolomite} = \frac{\text{Molecular weight of CaCO}_3}{\text{Molecular weight of dolomite}} \times 100$$

=100/92 Ã—100=108.7

Neutralizing value or CCE of some liming materials is given in table.

Table: Neutralizing value of some liming materials.

Liming materials	Neutralizing value of CCE (%)
Calcium oxide (Cao)	179
Calcium hydroxide $[Ca(OH)_2]$	136
Dolomite $[CaMg(Mg(CO_3)_2]$	108.7
Calcite $(CaCO_3)$	100
Basic slag $(CaSiO\ )$	86

(2) Purity of Liming Materials: The more purer the liming material, the higher will be its effectiveness for the amelioration of soil acidity.

(3) Degree of Fineness of Liming Materials: The rate of reaction of liming materials with an acid soil depends upon its fineness because finer materials increase the surface contact with the soil.

If the liming materials are coarse, the reaction will be slight. The magnitude of such reaction as affected by the fineness of liming materials can be evaluated by measuring the change of soil pH. The amount of finer fraction of liming materials will be required much less as compared to coarser fractions of the material to achieve a certain pH.

The fineness is measured in terms of the ability of a material to pass through a sieve having 60 holes of equal size in one linear inch. Such a sieve is called a 60 mesh sieve and a material passing through such a sieve is allotted 100% efficiency rating.

One such scale used in Ohio (USA) is given below:

	Efficiency of rating (%)
1. Material passing through a 60 mesh size	100
2. Material passing through a 20 mesh but not a 60 mesh sieve	60
3. Material passing through an 8 mesh but not a 20 mesh sieve	20

Percent effective Calcium Carbonate (ECC) or (Neutralizing Index)

The effective calcium carbonate (ECC) rating of a limestone or liming materials is one product of its calcium carbonate equivalent (CCE) and the fineness factor. The fineness factor is the sum of the product of the percentage of material in each of the three size fractions multiplied by the appropriate effectiveness factor.

Percent ECC or N.I. = CCE Ã — fineness factor.

Chemical Reactions between Liming Materials and Acid Soils during the Process of Amelioration:

(i) Oxides of lime (CaO): When oxides of lime like (CaO) are applied to an acid soil, it reacts almost immediately as follows:

$$\text{soil}\left]\begin{matrix}H\\H\\H\\Al\\Al\end{matrix}\right. + H_2O + 2CaO \rightarrow \left.\begin{matrix}H\\Ca\\Ca\\Al\\H\end{matrix}\right[\text{soil} + Al(OH)_3$$

(ii) Hydroxides of lime $Ca(OH)_2$: When hydroxides of lime like $Ca(OH)_2$ is applied for the reclamation of an acid soil, the following chemical reaction takes place:

$$\text{soil}\left]\begin{matrix}H\\H\\H\\Al\end{matrix}\right. + H_2O + 2CaO \rightarrow \left.\begin{matrix}H\\Ca\\Ca\\H\end{matrix}\right[\text{soil} + Al(OH)_3 + H_2O$$

(iii) Carbonates of lime ($CaCO_3$): When carbonates of lime like calcite is applied to an acid soil, a part of $CaCO_3$ undergoes solution and combines with H_2CO_3 to form soluble $Ca(HCO_3)_2$.

This calcium bicarbonate in solution form reacts with the soil colloids with the evolution of CO_2 as follows:

$$CaCO_3 + H_2CO_3 = Ca(HCO_3)_2$$

$$Ca(HCO_3)_2 + {}^{H}_{H}[Soil + 2H_2O + 2CO_2$$

And the rest portion of the limestone goes in close contact with the soil colloids in solid condition as follows:

$${}^{H}_{H}[Soil + CaCO_3 = Ca[Soil + 2H_2o + CO_2$$

Solid condition

(iv) Basic slag: Basic slag can also be used as liming material when it is applied to an acid soil the following chemical reaction takes place.

$$Soil]^{H}_{Al} + 2CaSiO_3 = 3H_2O = {}^{Ca}_{Ca}[Soil + 2H_2SiO_3 + Al(OH)_3$$

(Metalisic acid)

The metasilicic acid is weakly dissociated, much less so that the clay absorbed hydrogen (H^+) ions and the pH of the soil is raised.

Besides these chemical amendments, some other management practices have been found to be effective for the amelioration of an acid soil namely:

1. Use of basic fertilizers like sodium nitrate, basic slag etc.
2. Reducing leaching of basic cations by following appropriate soil and water management practices.
3. Acid tolerant crops like rice, potato, tea wheat, sweet potato, maize, brinjal etc. should be cultivated.

Lime Requirement and Liming Factor

Once a soil is found to be acidic and then it requires some management practices either chemical amendment with some liming materials or some cultural management practices in order to improve its fertility status through the modification of soil reaction to a favourable position.

It has been found that the correction of soil acidity as well as improvement of soil fertility through liming is most effective. The different intensity of soil acidity required different amount of lime. So determination of lime requirement from an acid soil is prerequisite in a liming programme.

The desirable soil pH range for most of the field crop is 6.0 to 7.0. The amount of lime required to be added to acidic soil to raise the pH of that soil to a desired value is known as lime requirement. Shoemaker buffer method is used for the determination of lime requirement of an acid soil.

Buffer method for determining lime requirement of soils appreciable amount of extracted aluminium. Lime requirement in terms of pure calcium carbonate can be observed from the table.

Table: Lime Requirement of an Acid Soil.

pH of soil buffer suspension (field soil sample)	Lime required to bring pH down to indicated level $CaCO_3$ In tonners per acre		
	pH 6.0	pH 6.4	pH 6.8
6.7	1.0	1.2	1.4
6.6	1.4	1.7	1.9
6.5	1.8	2.2	2.5
6.4	2.3	2.7	3.1
6.3	2.7	3.2	3.7
6.2	3.1	3.7	4.2
6.1	3.5	4.2	4.8
6.0	3.9	4.7	5.4
5.9	4.4	5.2	6.0
5.8	4.8	5.7	6.5
5.7	5.2	6.2	7.1
5.6	5.6	6.7	7.7
5.5	6.0	7.2	8.3
5.4	6.5	7.7	8.9
5.3	6.9	8.2	9.4
5.2	7.4	8.4	10.0
5.1	7.8	9.1	10.6
5.0	8.2	9.6	11.2
4.9	8.6	10.1	11.8
4.8	9.1	10.6	12.4

Liming Factor

'Liming factor' may be defined as the factor by which the actual amount of lime can be calculated from the estimated theoretical amount of lime. This factor varies from 1 to 3, depending on rate of limestone solution, plant uptake and leaching during the reaction period. A liming factor of 1.5 to 2.0 is generally used when the theoretical amount of lime necessary to bring the soil to a given pH is converted to the actual amount of limestone to be added in the field.

Methods of Applying Lime

The application of small amounts of lime of soil in every year or twice in a year has been found to be most effective. But this involves the application cost of the liming material. Generally lime should be applied well ahead of the crop cultivation and broadcasted limes are to be well mixed with the whole plough layer soil so that liming reaction can occur in a faster rate.

The usual liming practice consists of a compromise between what is most effective and what is cheapest per ton of lime applied. When both surface and sub-surface soils are strongly acidic e.g. ultisols, it sometimes pays to incorporate lime to a depth of about 30 cm (12 inches). Application of lime of no till field soils may not be as effective as application to an equivalent cultivated soil.

Lime Balance Sheet

When a soil has had its acidity corrected by lime, how often must lime be added and how much is needed to keep the soil pH suitable? The answer depends upon the rate of lime loss.

Lime is neutralized or lost from the soil by six activities:

- Neutralization by acid-forming fertilizers, a rapid change.

- Neutralization by the acid formed by CO2 in water (from organic matter decomposition), a slow process.
- Leaching a relatively slow change.
- Removal in harvested or grazed crops; relatively slow loss.
- Erosion, a top soil is lost with its higher base saturation and often leaving more acidic sub-soil to be limed.
- Neutralization by acids dissolved in precipitation, resulting from oxides of sulphur fumes from manufacturing plants, a slow process.

Effect of Over Liming

When excessively large amounts of lime applied to an acidic soil the growth of plants is affected by influencing either one or many of these following causes:

- Deficiency of iron, copper and zinc will occur.
- Phosphorus and potassium availability will be reduced.
- Due to high OH^- ion concentration by over liming, root development will be inhibited in association with tip swelling brought about by hydrations. Due to dehydrating properties of boron, it acts as a protective agent for excess OH^- ion concentration.
- Due to over liming, boron deficiency will occur.

$$2Alx_3 + 3CaCO_3 + 3H_2O = 3Cax_2 + \underset{\text{(Precipition)}}{2Al(OH)_3 \downarrow} + 3CO_2$$

  where 'x' is the exchange site.

  The freshly $Al(OH)_3$ is then available for adsorption of boron. This reaction mechanism is involved for causing boron deficiency.

- Due to application of lime in excess, the incidence of diseases like scab in root crops will be increased.

All these effects can be reduced with the application of large amounts of organic manures like well rotten farm yard manure, green manure crops, compost, phosphorus, boron or a mixture of micronutrient fertilizers to the soil.

Exchangeable Ions in Soil

Soil colloids are the seat of reaction. Ion exchange (cation and anions) takes place in colloids. The phenomenon of ion exchange is of great importance in agriculture. It has considerable influence on the liberation of plant nutrients such as Ca, K, P etc. It controls soil structure and crumb formation. It is also responsible for imparting a stable structure. It controls the processes plays an important role in the reclamation of acid and alkali soils. It also influences the effect of fertilizers and fertilizer practices.

Ion exchange is two types: cation exchange or base exchange and anion exchange or acid exchange. Ion exchange is a reversible process in which cation and anion exchanged between solid and liquid phase.

Cation Exchange or Base Exchange

In a near neutral soil, Ca is remaining adsorbed on colloidal particle H ion generated as organic and mineral acid formed due to decomposition of organic matter. In colloid, H ion is adsorbed more strongly than is the Ca and H is chemically equivalent.

The reaction is as follow:

$$Ca - Clay\ Micelle + 2H + \rightleftarrows {}^{H}_{H} > Clay\ Micelle + Ca^{2+}.$$

This phenomenon of the exchange of cations between soil and salt solution is known as calion exchange or Base Exchange and the cations that take part in this reaction are called exchangeable cations. It is the interchange between a cation in solution and another cation on the surface of clay or organic matter (colloids). Thompson and Way first recognised cation exchange phenomenon. They passed a solution of $(NH_4)_2SO_4$ through a soil column and found that effluent contained no NH_4^- salt but $CaSO_4$. This phenomenon is always in equivalent quantities of ions.

$$\begin{array}{cc} Ca^{+2} & Ca^{+2} \\ \multicolumn{2}{c}{\boxed{\text{Soil Colloids}}} \\ Ca^{+2} & Ca^{+2} \end{array} + (NH_4)_2SO_4 \rightleftarrows \begin{array}{cc} NH_4^+ & NH_4^+ \\ \multicolumn{2}{c}{\boxed{\text{Soil Colloids}}} \\ NH_4^+ & NH_4^+ \end{array} + CaSO_4$$

Types of Silicate Clay Mineral

In case of kaolinite, monovalent cations are more strongly held than divalent, while in case of montmorillonite, divalent cations are strongly held than monovalent cations. The normal order of adsorbed cations on clay particles is:

$$H^+ > Al^{+3} > Ca^{+2} = Mg^{+2} > K^+ = NH > Na^+.$$

Significance of Cation Exchange

Cation exchange is an important reation in soil fertility, in causing and correcting soil acidity and basicity. In changes altering soil physical properties, and as a mechanism in purifying or altering percolating Waters. The plant nutrients like calcium, magnesium and potassium are supplied to plants in large measure from exchangeable forms.

Cation exchange is very important in soils because of the following relationships:

- The exchangeable K is a major source of plant K.

- The exchangeable MG is often, a major source of plant MG.
- The amount of lime required to raise the pH of an acidic soil is greater as the CEC is greater.
- Cation exchange sites hold Ca^{+2}, Mg^{+2}, K^+, Na^+ and NH_4^+ ions and slow down their losses by leaching.
- Cation exchange sites hold- fertilizer K^+ and NH_4^+ and greatly reduces their mobility in soils.
- Cation exchange sites adsorb many metals (Cd^{+2}, Zn^{+2}, Ni^{+2} and Pb^{+2}) that might be present in wastewater. Adsorption removes them from the percolating water, thereby cleansing the water that drains into groundwater.

Anion Exchange

The process of anion exchange is similar to that of cation exchange. Under certain conditions hydrous oxides of iron and aluminium shows evidence of having positive charges on their crystal surfaces. The positive charges of colloids are due to addition of hydrogen (H^+) in hydroxyl group (OH^-) resulted in net positive charge (OH^{2+}) This positive charge will attract anions. The capacity for holding anions increases with increase in acidity. The lower the pH the greater is the adsorption. All anions are not adsorbed equally readily. Some anions such as H_2PO_4 are adsorbed very quickly at all pH values in the acid as well as alkaline range Cl^- and SO_4^- ions are adsorbed slightly at low pH but none at neutral soil, while NO_3^- ions are not adsorbed at all. Hence, at the pH commonly prevailing in cultivated soils- nitrate NO_3^- chloride (Cl^-) and sulphate (SO_4^-) ions are easily lost by leaching. In general, the relative order of anion exchange is:

$$OH > H_2PO_4^- > SO_4^- > NO_3^-$$

$$R.OH + HCI\, R.OH_2 + Cr$$
$$R.NH + HCI\, R.NH_3 + Cr$$

This adsorbed negatively changed anion is replaceable as under:

$$\underset{(R=Al.Fe)}{R.OH^+_2 +} \quad \underset{\text{soil solution}}{Cr + H_2PO_4^-} \rightarrow \underset{\text{on exchange complex}}{R.OH_2.H_2} \quad \underset{\text{soil solution}}{+\ Cr}$$

In this way there is fixation of phosphate ion, in which OH ion of silicate clay mineral is substitute by H_2PO_4 ion by isomorphrous substitution. The H_2PO_4 ion becomes a part of the silicate clay minerals and the phosphate fixed by this way is known as colloid bound phosphate. This phosphate is not exchangeable and available to the plant. The anions like NO_3 and OH are not bound at pH above 7. The adsorbing power of anions is in the order of:

$$Cl^* = NO_3{}^* < SO_4 < PO_4$$

*are not held by the soil colloids and are lost due to the leaching.

$$or\, OH > H_2PO_4 > SO_4 > NO_3$$

Importance of Anion Exchange in Nutrient Availability

The phenomenon of anion exchange assumes in relation to phosphate ions and their fixation. The exchange is brought mainly by the replacement of OH ions of the clay minerals.

$$\text{OH Clay} + \underset{\text{Available}}{H_2PO_4} \rightarrow \underset{\text{Unavailble}}{H_2PO_4\,\text{Clay}} + \text{OH}$$

The adsorption of phosphate ions by clay particles from soil solution reduces its availability to plants. This is also known as colloid bound phosphate fixation. The phosphate ion again becomes available when lime is applied to increase the pH of acidic soil.

The OH ion originated not only from silicate clay minerals but also from hydrous oxides of iron and aluminum present in the soil. The phosphates ions react with the hydrous oxides also get fixed forming insoluble hydroxyl phosphate of ion and aluminum. Which is called saloid bound phosphate.

$$Al(OH)_3 + \underset{\text{Available}}{H_2PO_4^-} \rightarrow \underset{\text{Unavailble}}{Al(OH)_2 + H_2PO_4} + OH^-$$

If the reaction takes place at a low pH under strongly acid conditions, the phosphate ions are irreversibly fixed and are totally unavailable for the use of plants.

Cation Exchange Capacity (CEC)

Cations are positively charged ions such as calcium ($Ca2^+$), magnesium ($Mg2^+$), and potassium (K^+), sodium (Na^+) hydrogen (H^+), aluminum (Al^{3+}), iron (Fe^{2+}), manganese (Mn^{2+}), zinc (Zn^{2+}) and copper (Cu^{2+}). The capacity of the soil to hold on to these cations called the cation exchange capacity (CEC). These cations are held by the negatively charged clay and organic matter particles in the soil through electrostatic forces (negative soil particles attract the positive cations). Thecations on the CEC of the soil particles are easily exchangeable with other cations and as a result, they are plant available. Thus, the CEC of a soil represents the total amount of exchangeable cations that the soil can adsorb.

The cations used by plants in the largest amounts are calcium, magnesium, and potassium. In most soils within humid regions such as in New York, sodium is not present in sufficient quantities to occupy a significant amount of the CEC. However, in dry climates, sodium can occupy an important portion of the CEC. Other cations that can occupy cation exchange sites in New York soils are hydrogen, aluminum, iron and manganese. Cations such as zinc and copper are typically present in the soil in too low a concentration to occupy much of the CEC.

Why do Soils have a CEC?

Soils have a CEC primarily because clay particles and organic matter in the soil tends to be negatively charged. New York soils have silicate clay minerals (clay minerals that contain silica). Each silicate clay particle is made up of individual layers or "sheets". If the mineral was pure silica and oxygen (silicaoxide more commonly referred to as quartz), the particle would not have any charge. However, clay minerals common in New York agricultural soils, contain aluminum as well as silica. They have a net negative charge because of the substitution of silica (Si^{4+}) by aluminum (Al^{3+})

in the mineral structure of the clay. This replacement of silica by aluminum in the clay mineral's structure is calledisomorphous substitution, and the result is clays with negative surface charge.

Since the soil as a whole does not have electric charge, the negative charge of the clay particles is balanced by the positive charge of the cations in the soil. The negative charges associated with isomorphous substitution are considered permanent, that is, the charges do not change with pH changes.

Mineral Soil Particles

Sand Clay

S_i2O_4 $SiAlO_4^-$

No charge.
Does not retain
cations.

Negative charge
Attracts
and retains cations.

Substitution of silica by aluminum in soil clay particles causes clays to have a negative charge. Because of this negative charge, the soil can hold on to positively charged cations such as calcium (Ca^{2+}), magnesium (Mg^{2+}) and potassium (K^+)

Organic matter can have a 4 to 50 times higher CEC per given weight than clay. The source of negative charge in organic matter is different from that of clay minerals; the dissociation (separation into smaller units) of organic acids causes a net negative charge in soil organic matter, and again this negative charge is balanced by cations in the soil. Because organic acid dissociation depends on the soil pH, the CEC associated with soil organic matter is called pH-dependent CEC. This means that the actual CEC of the soil will depend on the pH of the soil. Given the same amount and type of organic matter, a neutral soil (pH ~7) will have a higher CEC than a soil with e.g. pH 5, or in other words, the CEC of a soil with pH-dependent charge will increase with an increase in pH.

Units

The CEC of a soil is expressed in $cmol_c$/kg (centimol positive charge per kg of soil) or meq/100 g (milli-equivalents per 100 grams of soil). Both expressions are numerically identical (10 $cmol_c$/kg = 10 meq/100 g).

Common CEC Ranges

Sandy soils low in organic matter have a very low CEC (less than 3 $cmol_c$/kg) while heavier clay soils or soils high in organic matter generally have a much higher CEC (greater than 20 $cmol_c$/kg). Table gives an estimate of a typical CEC of soils of each of the five soil management groups in New York.

Table: Estimated cation exchange capacity (CEC) of soils typical for New York State agricultural land.

SMG*	General Description	CEC ($cmol_c$ /kg soil)
1	Fine-textured soils developed from clayey lake sediments and medium- to fine-textured soils developed from lake sediments.	25

2	Medium- to fine-textured soils developed from calcareous glacial till, medium-textured to moderately fine-textured soils developed from slightly calcareous glacial till mixed with shale, and medium-textured soils developed in recent alluvium.	20
3	Moderately coarse textured soil developed from glacial outwash and recent alluvium and medium textured acid soil developed on glacial till.	18
4	Coarse- to medium-textured soils formed from glacial till or glacial outwash.	16
5	Coarse- to very coarse-textured soils formed from gravelly or sandy glacial outwash or glacial lake beach ridges or deltas.	12

For New York soils, it is better to *measure* the CEC rather than use the values reported in table, as the CEC will change with soil pH (i.e. is pH dependent). Soil pH changes can be caused by natural processes, such as decomposition of organic matter and leaching of cations, but also by human actions such as application of acidifying nitrogen fertilizers and liming materials. Most laboratories approximate the CEC from regular soil test results. So, the CEC value listed on regular soil test report is the results of a *calculation*, not an actual measurement.

Implications

- The higher the CEC the more clay or organic matter present in the soil. This usually means that high CEC (clay) soils have a greater water holding capacity than low CEC (sandy) soils.
- Low CEC soils are more likely to develop potassium and magnesium (and other cation) deficiencies, while high CEC soils are less susceptible to leaching losses of these cations. So, for sandy soils, a large one-time addition of cations e.g. potassium can lead to large leaching losses (soil isn't able to hold on to the excess K). More frequent additions of smaller amounts are better.
- The lower the CEC, the faster the soil pH will decrease with time. So, sandy soils need to be limed more often than clay soils.
- The higher the CEC, the larger the quantity of lime that must be added to increase the soil pH; sandy soils need less lime than clay soils to increase the pH to desired levels.

Mineral Absorption by Plants

There are 118 or so different elements in this planet, (there are 80 elements have at least one stable isotope and 38 of them are radioactive). Among the said elements $Fe^{+}(2^{+})$ is found as the most abundant element on this planet. Hydrogen and N_2 are the most abundant elements in atmosphere. Such elements which are absolutely required for the normal growth and development of the plant body are called essential nutrients. of which fourteen or fifteen are absolutely required for the life processes without which plants exhibit diseased symptoms and ultimately die. Among them, eight elements are required in sufficient quantities, others in small quantities; the former referred to as macronutrients and the latter as micronutrients or true elements. Nevertheless, plants also contain elements other than the elements mentioned above, whose deficiency may not cause any diseased symptoms or death and such elements are named as non essential elements. The macronutrients are carbon, hydrogen, nitrogen, phosphorus, potassium, calcium, sulphur, magnesium and iron. The micronutrients are manganese, zinc, boron, copper, molybdenum and cobalt. Non-essential elements are sodium, aluminum, silicon, chlorine, gallium, and others.

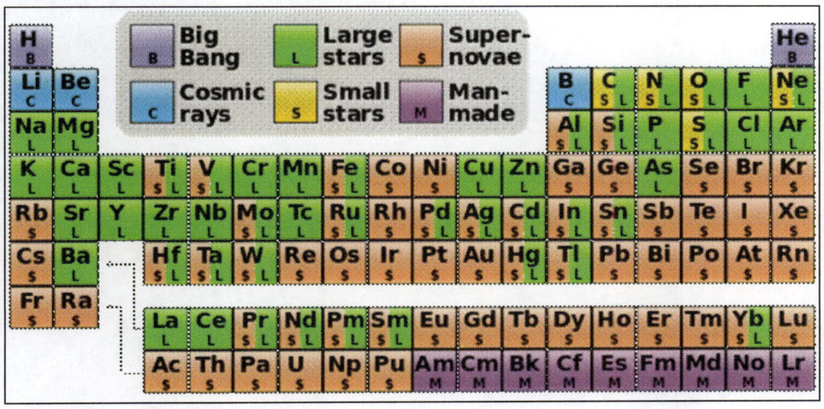

Periodic table showing the cosmogenic origin of each element in the Big Bang, or in large or small stars; Small stars can produce certain elements up to sulfur, by the alpha process. Supernovae are needed to produce "heavy" elements (those beyond iron and nickel) rapidly by neutron buildup, in the r-process. Certain large stars slowly produce other elements heavier than iron, in the s-process; these may then be blown into space in the off-gassing of planetary nebulae.

The periodical shown shows elements with increased atomic number showing the number of protons.; Elements as groups found in columns share recurring physical and chemical properties; Carbon found in the diamond is called allotrope.

Nutritional elements in the periodic table are- H,B,C,N,O,F; Li; Na, Mg, Si, P,S,Cl; K,Ca, V,Cr,Mn,Fe,Co,Ni,Cu,Zn, As,Sc,Br; Mo and I; The said elements are considered as nutrient elements in the said periodic table.

Most of the above said elements are found in soil solution either in the form of inorganic or organic salts or ions; which may exist in either in free-state or bound to clay particles. Whenever there is depletion of any free ions from the soil solution, respective ions are released from clay particles into the soil solution to maintain the equilibrium. This is achieved by a process called ion exchange process. This may be due to contact ion exchange mechanism or by carbonic acid ion exchange ion mechanism.

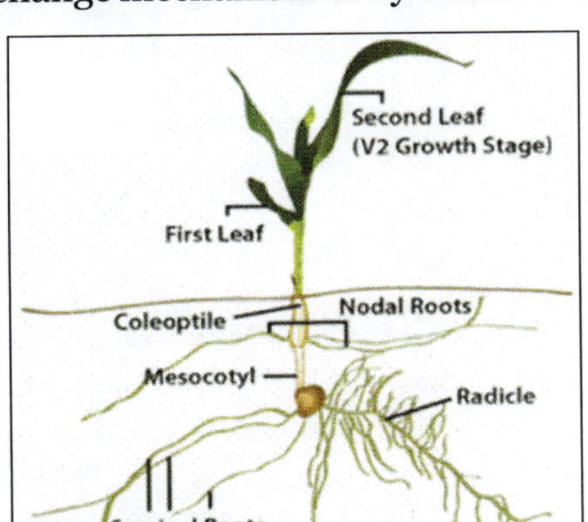

Aquatic plants do not need any special structures for the absorption of minerals, for the entire plant body acts as absorptive surface. But terrestrial plants possess extensive root system with innumerable growing apices. Short term radioactive isotope labeling experiments indicate that the meristematic regions of the root absorbs greater amount of ions than any other regions. This is perhaps necessitated by their active metabolic state. Though most of the minerals are absorbed by the growing meristems, minerals ultimately find their way into xylem elements by active transport. From the xylem elements they move upwards along with transpiration stream and spread with branches and sub branches thus get distributed to all other regions.

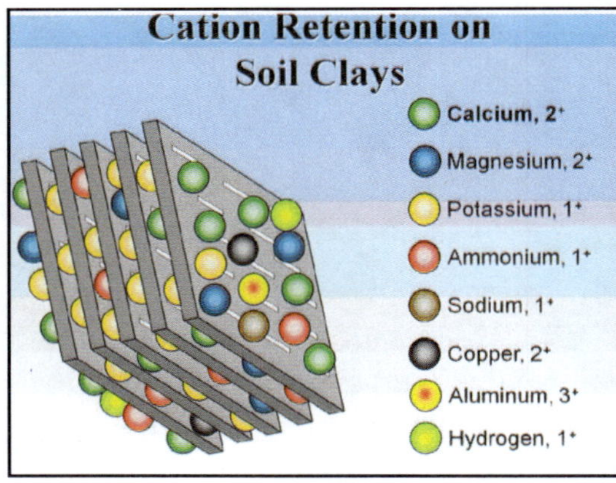

Clay particles have negative charges on their surfaces; so metal ions with positive charges bind to such negative charges.

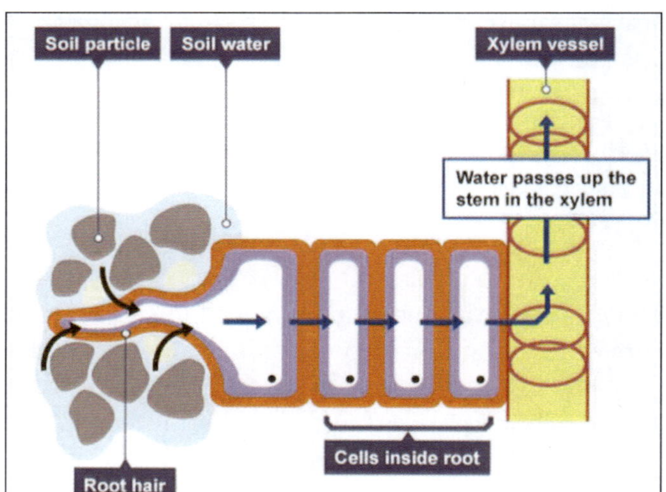

The pathway of water across a root

Root hair cells help to increase water absorption and other minerals. In order for water to be transported via the xylem tube, it moves from root cell to root cell via osmosis and apoplastic.

As water moves into the root hair cell down the concentration gradient, the solution inside the root hair cell becomes more dilute. This means that there is now a concentration gradient between the root hair cell and adjacent root cells, so water moves from the root hair cell and into the adjacent cells, till it reaches the xylem tube for transport, via osmosis. Mineral ions also move into apparent free spaces into roots, but to enter into cellular cytoplasm it requires different modes.

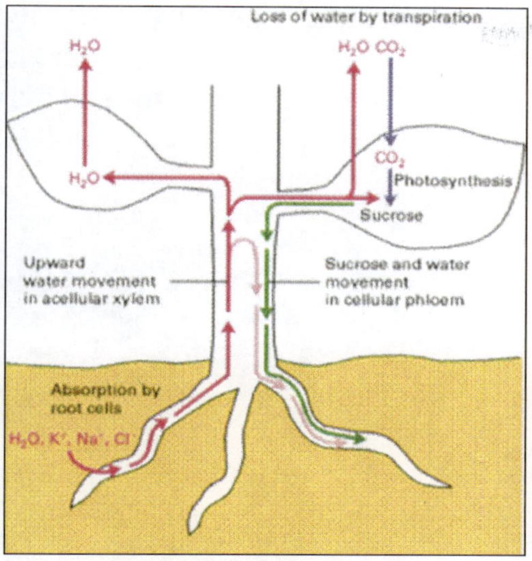

Mechanism of Absorption

Ion Exchange Mechanism

Plant roots are in contact with soil clay particulates which have colloidal dimensions. Most of the ions such as Na, K, and other are bound to colloidal soil particles. Root cells secrete hydrogen ions which are positively charged. Such ions can easily displace cat ions like K^+, Na^+ ions that are bound to clay particulates. Thus the cat ions are made available for root system to absorb the required ion. But the relative retentive capacity of the clay particles, though not absolutely fixed, is in the order of H^+>Ca^{2+}>Ng^{2+}>K^+>NH^{4+}>Na^+. But hydrogen ions are capable of replacing any cat ion found on such clay particles. Nonetheless it appears that hydrogen ions can replace any bound ions to clay particles but there is a preference of ions to be released. As hydrogen ions have greater affinity, they can replace any of the above ions easily.

Both cat ions and anions have a tendency to get adsorbed on to the surfaces of the cell walls, and exchange with ions present in the soil solution. This process of exchange between the adsorbed ions and ions in solution is known as ion exchange. Environmental factors operating in soil and cellular components affect/effect mineral absorption.

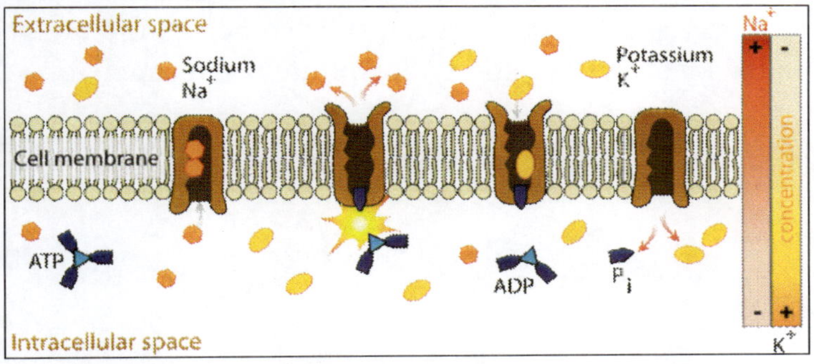

Plants utilize inorganic nutrients in the environment to produce essential metabolites. Nutrients effect efficiency thus increases plant fitness and yield performance. Acquisition of nitrogen (N) and sulfur (S) is particularly important since these elements are required for synthesizing amino acids and proteins. Nitrate (NO_3^-), ammonium (NH_4^+) and sulfate (SO_4^{2-}) are the major forms of N and S utilized by plants in nature. However, these nutrient resources are not uniformly distributed in the soil environment. Despite the application of fertilizers, the soil nutrient concentrations fluctuate due to climate changes.

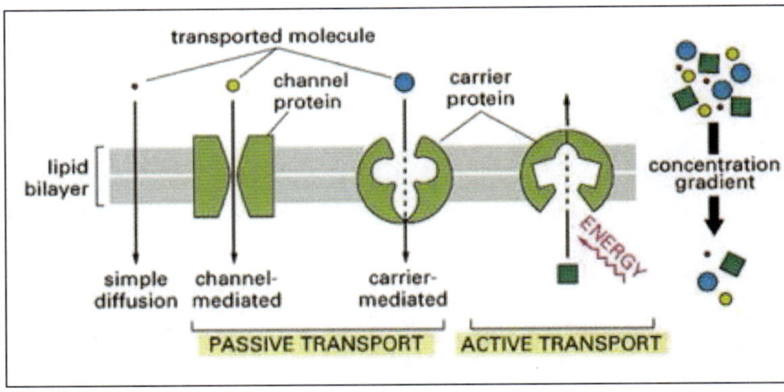

Essential Cell Biology

Once ions diffuse into cell surface of the roots, they cross into the inert cell wall and reach plasma membrane; there one finds different transporters some of which transport passively and some are transported across the membrane actively (use ATP energy).

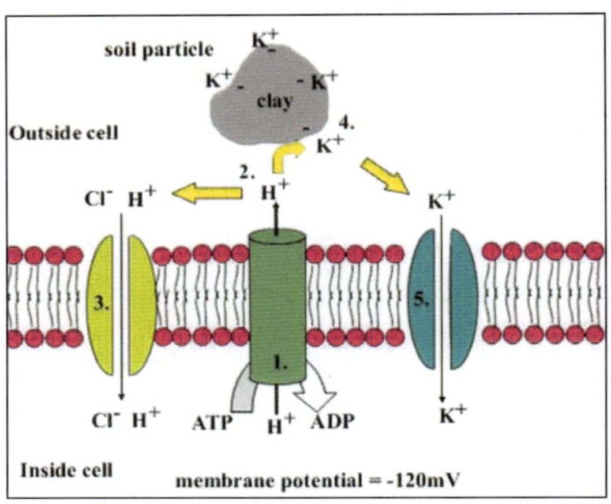

Cytochrome Pump- Salt Respiration or Electron Transport; This theory was proposed by H. Lundegardh, who suggested that anions could be transported across the membrane by cytochrome system. Energy is supplied by direct oxidation of respiratory intermediate.

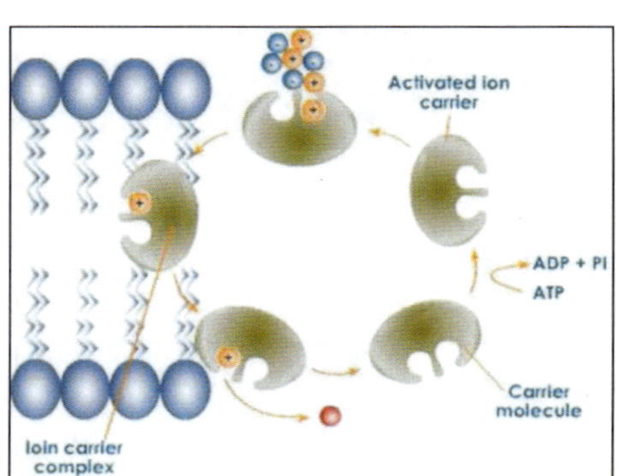

Within the cytoplasm, the complex breaks to release the ions. The carrier moves out of the cytoplasm and is again ready to attach another ion to from a complex.

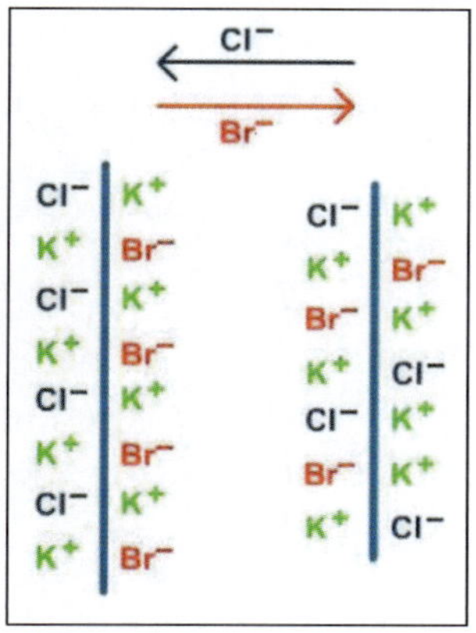

Negatively charged Cl^- and Br^- are exchanged without disturbing the electrical neutrality; this theory describes the effect of fixed or non-diffusible ions which mostly accumulate on the inner surface of the outer membrane. Process is named after its discoverer F.G. Donnan. If there is a negative non-diffusing charge on one side of a membrane, it will create a potential gradient across the membrane from which ions will diffuse. The result will be an electrochemical equilibrium. The concentration (chemical potential) of ions will not necessarily be the same inside and outside. Thus, as an electrical disequilibrium is maintained because of diffusing charges thus concentration disequilibrium is established.

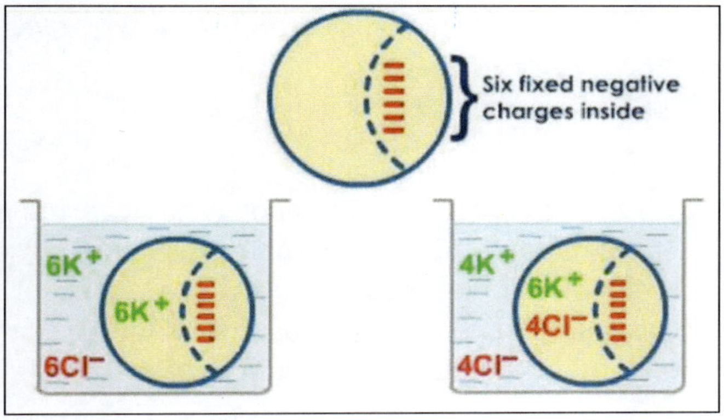

$$\frac{\text{Positive ions inside}}{\text{Positive ions outside}} = \frac{\text{Negative ions outside}}{\text{Negative ions inside}}$$

Therefore, according to Donnan, Donnan equilibrium is attained if the product of anions and cations in the internal solution becomes equal to the product of anions and cations in the external solution.

Carbonic Acid Ion Exchange Mechanism

Roots continuously respire irrespective of day or night, and liberate significant quantities of CO_2, which when dissolved in soil water produces carbonic acids. Immediately they ionize into H^+ and bicarbonate ions (HCO_3).

The hydrogen ions thus formed are capable of exchanging with any bound cations on clay particles and make the bound cations available for the roots.

Roots liberate CO_2

$$CO_2 + H_2O. \rightarrow H_2CO_3 \rightarrow H^+ + HCO_3$$

$$HCO_3 \rightarrow \text{ ion exchange with cations; } NH_4, K^+, Ca^{2+}, Mg^{2+} \text{ etc}$$

$$HCO_3 \rightarrow \text{ Ion exchange with anions } -No_{3\_}, H_2PO\$\text{-} HPO_{4\_}, Cl^- \text{ etc.}$$

Though the above mechanisms were once proposed as theories, it is now clear that growing roots not only release CO_2 but also secrete hydrogen ions. Thus the root does provide hydrogen ions for both carbonic acid ion and contact ion exchange processes. Similarly adsorbed anions are also exchanged by anions like OH^- ions.

Factors that Control Absorption

Soil aeration: In most of the cases, living cells cannot survive without oxygen. As the roots contain a large number of living cells, they require considerable amount of energy for their metabolic activities and growth. So oxygen is absolutely essential for generating energy rich components by biological oxidative process. As mineral absorption requires energy, poor soil aeration affects the ability of roots to absorb adequate quantities of minerals. Water logged soils or soils with higher content of clay have very little amount of air; under such conditions roots are subjected to anaerobic conditions and the absorption of minerals is drastically affected. The inhibition of absorption of minerals due to the effect of respiratory poisons on roots clearly suggests that the absorption of minerals is an energy dependent process.

Temperature

Soil temperature has a significant effect on roots metabolic activities and also it affects the mobility of ions in soil solution. If the temperature of the soil is lowered, absorption of minerals will be drastically reduced; but with the increase in temperature, the rate of absorption also increases, but up to certain limits. Drastic variations in the rate of absorption due to changes in the temperature suggest, the process is dependent on protein or enzymatic activity.

pH of the Soil Solution

The degree of ionization of minerals and other nutrients depends upon the hydrogen ion concentration of the soil solution. For example, most of the phosphate ions, at alkaline pH exist either as bivalent H_3PO_4 ions or trivalent H_3PO_4 ions. Such ions are not favored for absorption. On the other hand neutral pH favors the absorption of monovalent ions. So the soil pH has a significant effect not only on the rate but also the kind of ions uptake. This property is also due to its effect on cellular components that are involved in absorption, which further suggests that proteins are involved in the ion uptake. As the protein structure is very sensitive to pH, its function also changes if there is any change in the pH. That is why the maintenance of proper soil pH is very important in agriculture. Too acidic or too alkaline soil is virtually useless for cultivation. Until and unless the soils are restored in terms of pH, such soils remain as wastelands.

Concentration of Soil Solution

Generally the concentration of minerals and its components found in soil solution is far below the levels of the same found in the cell sap. It means that the absorption of ions takes place against concentration gradient. The relative concentration of ions found in the cell sap and soil solution gives absorption ratio.

Ocean water contains relatively greater amount of salts than that of fresh waters. The land plants which are adapted to grow in fresh water soils die in marine water, because the marine water is enriched with greater amount of metal ions. Physiologically dry for them. But marine plant cells which have been adapted to such waters contain much more ionic contents than found in sea water. Even here, the ions are absorbed against concentration gradient.

The rate of absorption of ions very and depends upon the concentration of the soil solution. Normally, roots absorb greater amount of ions at a greater rate in dilute solutions than in a relatively high concentration solutions. How exactly the dilution enhances the rapid uptake is not clear, but it is a fact.

Ion-Ion Interactions

Soil solution consists of a wide variety of ions in different concentrations. While roots absorb inorganic nutrients, the ions of one kind present in the solution, either facilitate or interfere with the uptake of the other kind of ions. This phenomenon is called ion antagonism. On the other hand, a particular species of or ions enhance the uptake of another kind of ions. Such a phenomenon is referred to as ion facilitated uptake.

Epstein has demonstrated that the absorption of K and Fe is antagonized by the presence of calcium and magnesium bivalent ions. Similarly $CaCl_2$ has been found to inhibit the uptake of Cu_2 ions and save the plants from copper toxicity. On the other hand, sodium chloride has been found to facilitate the uptake a wide variety of ions.

Such type of ion interactions leading to antagonistic or facilitated uptake is explained on the basis of carrier molecules. Different ions have different carriers or transport proteins. Because of the specific binding site, any ion that competes with the other ion for the same sites results in ion antagonism. On the contrary, a particular ion binding to carrier molecules facilitates the binding of specific ion and enhance uptake of the said ion. So the balanced inorganic nutrient is very important, otherwise roots absorb more of one kind of ions or the absorption f an essential ion may be prevented by the presence of another kind of ion.

Important Features of Absorption

Unequal Absorption and Specificity of Ion

If a mixture of different elements of equal molar concentration in the form of a buffered solution is provided to the root system, it absorbs some ions in greater amounts than other, the rest are absorbed in traces with variations. This indicates the unequal uptake and also specificity. Certain cells, at a particular stage of development, absorb specific ions because they are required for their metabolism. The specificity is demanded by the needs of cells or tissues. In spite of it, the pH of the external solution remains more or less neutral. This is certainly due to exchange of ions. This can be demonstrated by placing a tomato plant with its roots intact in a dilute solution of NaCl. After a period of time, certain ions like K and Ca_2 are found in the external solution, which were not present before. This phenomenon explains the exchange of ions between the external solution and internal sap. Another equally important aspect of unequal uptake or absorption of ions is the dilution effect, where greater the dilution of external solution greater is the rate of uptake. This behavior is difficult to explain. Furthermore, the preferential uptake clearly suggests the role of specific carriers in the process of absorption of ions.

Salt Accumulation

Analysis of the concentration of specific nutrients present within the cell sap and the external solution reveals, that the relative concentration of specific components show greater concentration within the cells than in the external solution. Use of radioactive isotopes as traces also supports the same view, where certain ions are accumulated or taken in against concentration gradient. If the same is expressed in terms of chemical potential measured as milli volts. In Nietellas' cellular cytoplasm, Na^+ shows a chemical potential of 72 mV, potassium shows a difference of 40 MV and chorine +237mV. The above observations clearly suggest that the concentration of Na is very high in the external solution. In normal course, it should diffuse into the cell across the membrane by passive process, but to maintain chemical potential gradient and to prevent excess Na toxicity, Na ions are expelled out of the cell. On the contrary, the movement of CI is an uphill journey, because the concentration of them is many folds higher inside the cell than outside. The movement of ions inwards and outwards is referred to as ion flux and ion efflux respectively.

Saturation Effect

If a root system is provided with an excess amount of specific ions, initially ions are taken up at a greater rate but later the rate of uptake remains steady and constant. This observation further suggests that for a given ion there is a fixed number of specific carrier sites; if all are loaded with their respective ions, the rate of uptake can not be increased until and unless the number of carriers is increased.

Metabolic Energy

Energy is required for all metabolic processes. To demonstrate whether metabolic energy is required for ion uptake or not, it is possible to test it by providing respiratory poisons like KCN, DNP, rotenone, etc, to the root system engaged in the absorption of minerals. As soon as the inhibitors are added, the uptake of ions drastically reduces, which suggests that energy is absolutely required for the absorption of ions.

Apparent Free Space

If a plant, with its root system intact, is provided with a known amount of radioactive ions like 35 SO_4 or 32P for about 30 minutes, it is possible to determine the total amount of ions taken up by the root system by measuring the amount of radioactivity left in the exogenously provided solution.

Then if such a plant is transferred to water significant amount of radioactivity re appears in the water which can be easily quantified. The difference between the total amount of radioactivity absorbed during the first incubation and the total radioactivity that re appears in water is actually the amount of radioactivity taken into the root cells. The radioactivity that diffuses out into water is the amount of radioactive ions that taken into the free spaces found in the root cells. Thus, the free spaces are free for diffusion of ions and such spaces are called apparent free spaces. In roots such spaces are found in the intercellular regions and cell walls. In fact estimations show that the total AFS found is substantial in comparison to the total volume of the root. The movement of ions or water in AFS is a passive process and concentration dependent.

Donnan's Free Space

A physiologist by name Donnan proposed that all cells contain the some negatively charged macromolecules in the outer region of the cytoplasm. The space occupied by such molecules is called Donnan's space. In order to neutralize such charges, ion pairs like potassium and chloride enter into the cell in equal amounts. Some of the K^+ ions get adsorbed onto the negatively charged Donna's molecules. This leads to greater negative charge within the cell. So, to neutralize it, more of K^+ ions enter. Thus greater amount of K ions are absorbed. This process is believed to be passive. Furthermore, such spaces are assumed to be found within the peripheral region of the cytoplasm.

Some people have estimated that Donnan's molecules are supposed to be ureic acid molecules. In reality, most of the organic macromolecules found in the cytoplasm do show both charges. Then, where does the Donnan's space is found in the cell? It is difficult to explain ion uptake by Donnan's mechanism.

Mechanism of Absorption

From time to time, various theories have been proposed to explain the mechanism of ion uptake or absorption. Salt respiration theory by Lundergardh, cation ladder theory by Middleton and Russell, lecithin carrier hypothesis by Bennet and Clark and others have made attempts to explain the mechanism, but all of them fail to explain experimental observations. Though the said theories have no relevance to what we known today, these theories have been retained in many textbooks as historical landmarks in understanding the mechanism.

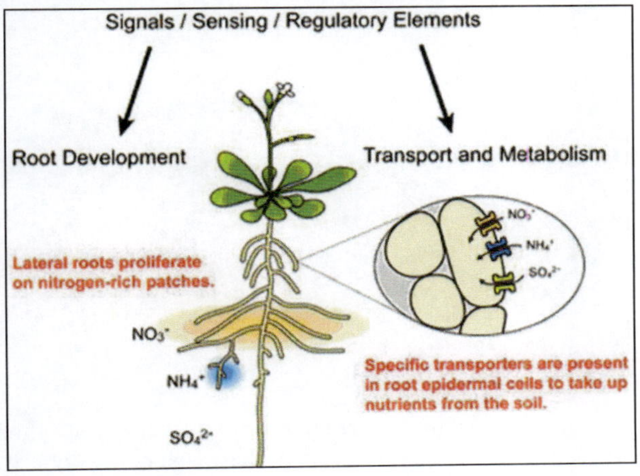

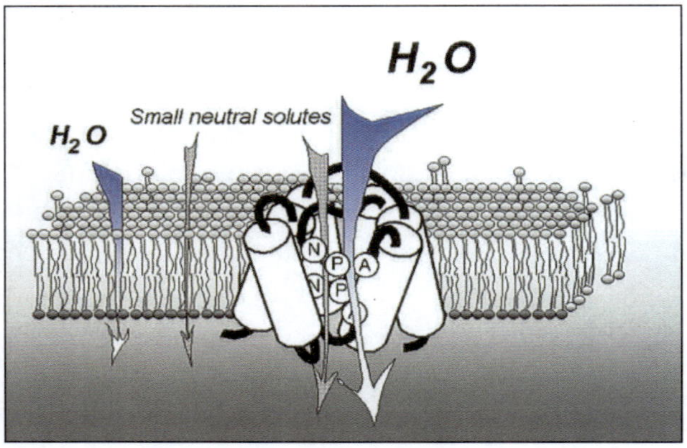

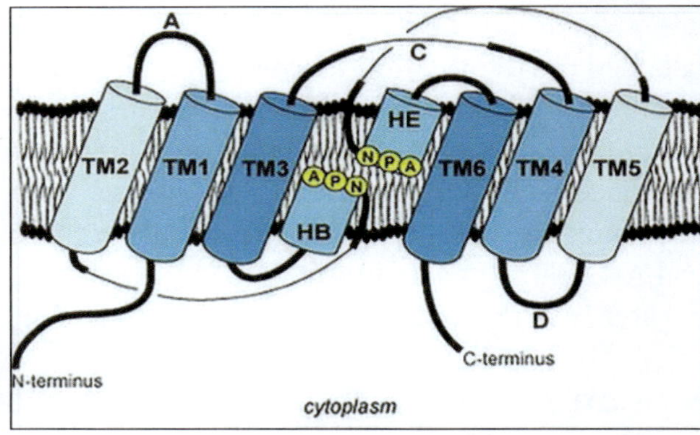

Aquaporin is a transmembrane multipass protein act as transporter of water in bulk. It also transports small neutral components

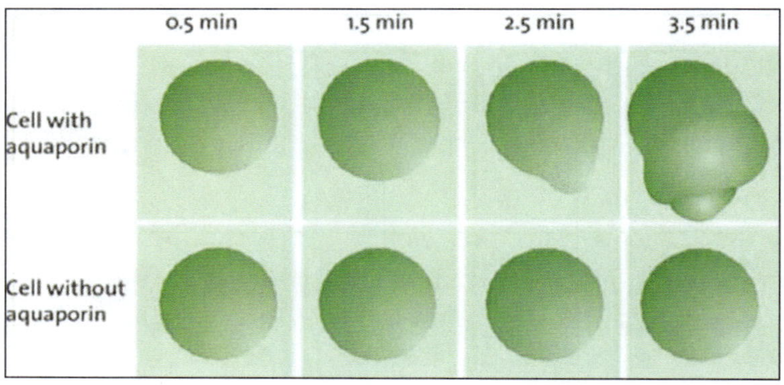

Aquaporins efficiency in water transport

As evidenced by experimental observations, the overall process is not a single mechanism, but it is a combined action of multiple events. The present explanation is based on the evidences obtained by using radioactive tracers, respiratory inhibitors, ionophores and mutants in plants, animals and microbes. The entire process of ion absorption takes place in two phases i.e. first passive phase and the second active phase. It is to be noted that plant absorb not just cat ions but also anions too.

Passive Phase

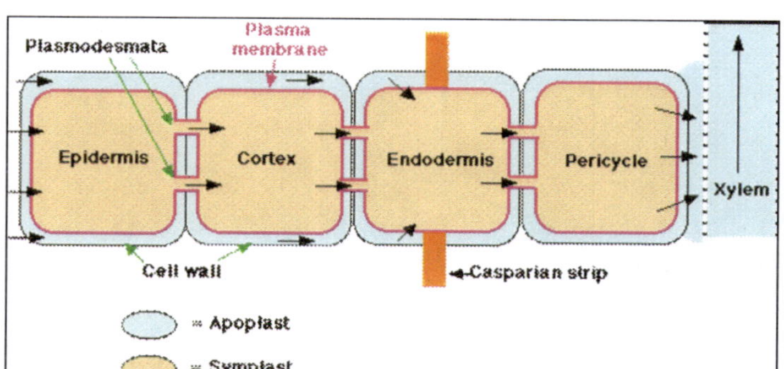

Minerals move into apoplasts epidermal and cortical cells then then the move into vascular xylem elements via passage cells in endodermis

When the intact root system is provided with an exogenous supply of mineral nutrients, to begin with, ions diffuse into the apparent free spaces found within the cell wall and the intercellular spaces found in the tissue. The rate of diffusion depends upon the steepness of the ionic gradient between the external solution and the solution found in AFS.

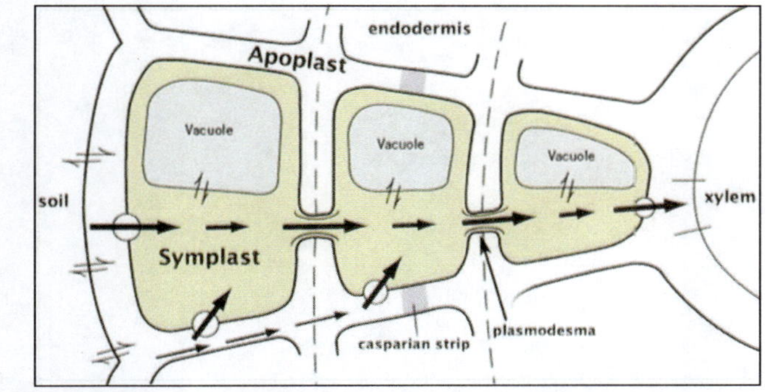

Apoplast-movement in extracellular space; symplastic-movement cytoplasm to cytoplasm by plasmodesmata

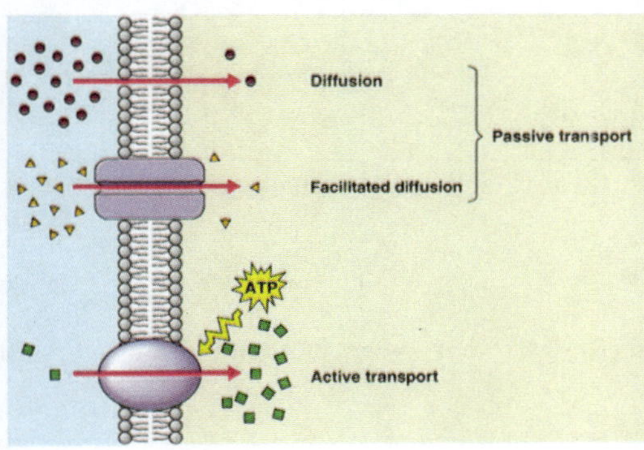

Proton pumps within the plasma membrane pump out H^+ ions into the soil. These H^+ ions combine with anions (Cl^-) that allow the uptake of the ions against the electrochemical gradient. H^+ also displaces K^+ from the clay particles in the soil, which allows them to travel through the electrochemical gradient through facilitated diffusion.

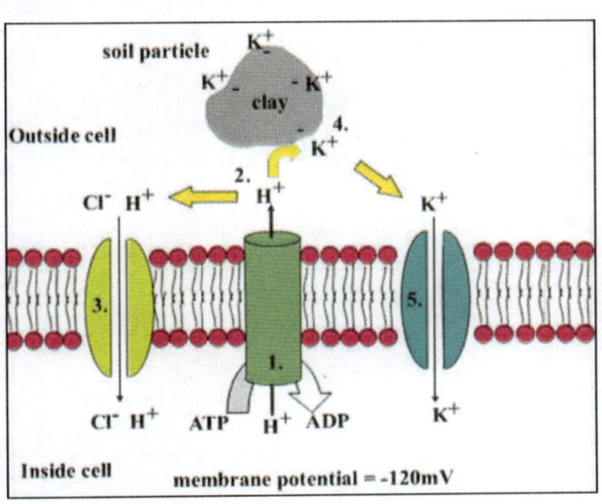

K^+ is present in the clay and soil that surround the root. They can be actively taken up by the active transport membrane pumps. Through active transport the mineral ions pass through and enter the cell.

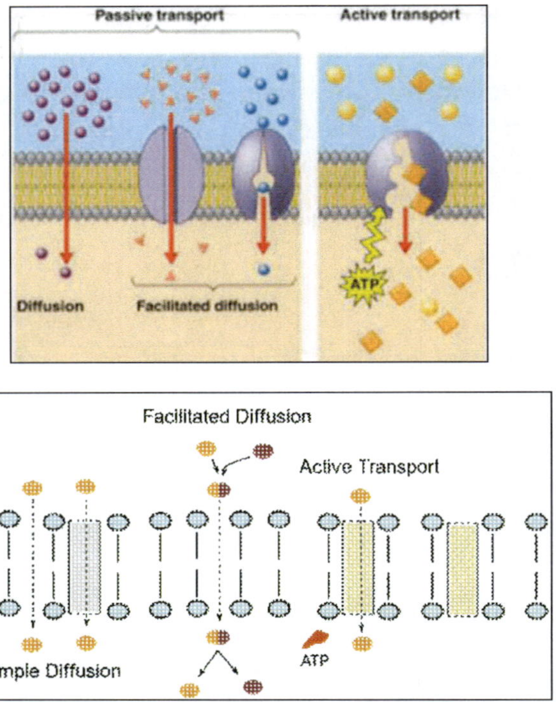

Diffusion, Facilitated Diffusion, Active Transport

Membrane Permeability

- Passive transport relies on diffusion to move particles across a membrane; down their concentration gradient.
- Active transport relies on energy consumption to move particles across a membrane, up their concentration gradient.
- Transport proteins can bind to solute and move them through the membrane while selective protein channels form pores through which only specific solutes can pass.

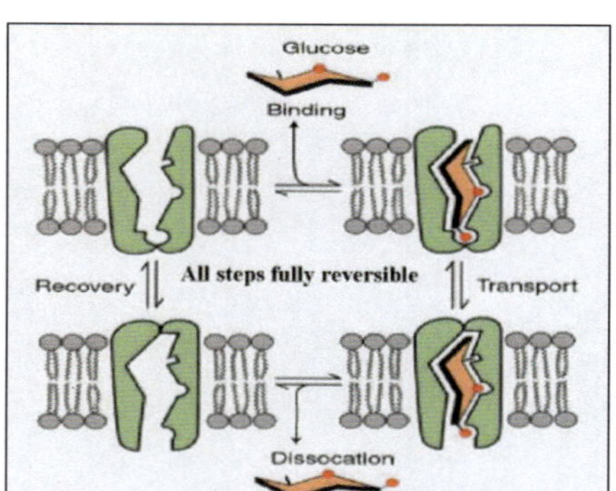

Mechanism for facilitated diffusion

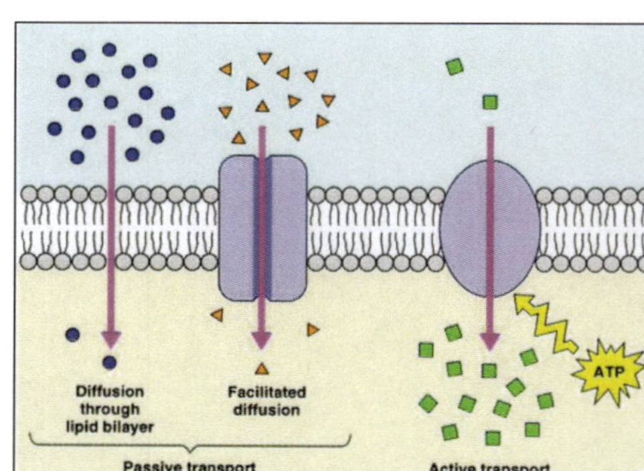

A comparison of passive and active transport

In fact the fine micro spaces found in the cell walls never act as an impediment for the free diffusion of ions, instead the charged cellulose material greatly facilitates rapid diffusion of ion into the cell wall spaces. Thus ions diffuse through the AFS of the root system and cross through the cortical cells and outer xylem elements through the passage cells. In xylem cells, the concentration of ions is always low because of continuous upward movement of sap by transpiration pull. So the initial uptake is very rapid and a passive process. Moreover this entire event is a kind of mass movement of ions along the concentration gradient till they reach the plasma membrane.

It is not just mineral uptake is facilitated by protein transporter; even water is carried by their own protein carriers called aquaporins. They have transcellular domains, at one surface they have H_2O binding site, when H_2O binds, it changes its structure, in such a way water is released at the other side.

Active Phase

Once the AFS are filled with mineral nutrients by free diffusion, the entry of ions across the plasma membrane into the cellular cytoplasm is highly regulated, in the sense the uptake is selective and energy dependent. Though some ions, because of local destabilization of membrane potential or due to membrane leakiness diffuse across the membrane along the concentration gradient, but the entry of majority of ions is selective, facilitated and dependent on metabolic energy. Probably that might be the reason as to why most of the ions absorbed by the root system are found in meristematic cells because they are actively metabolic and produce more of ATPs than any other cells.

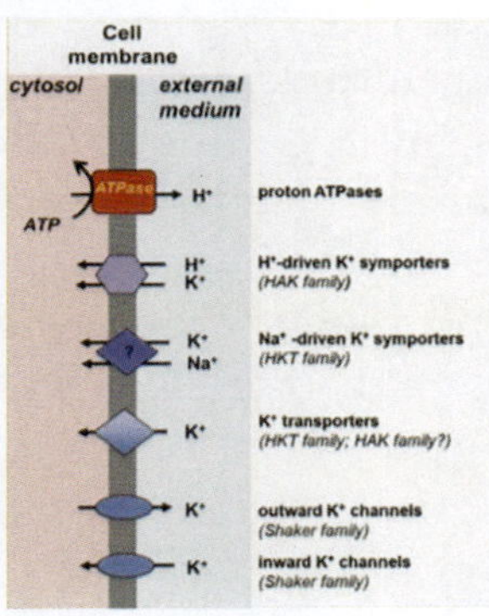

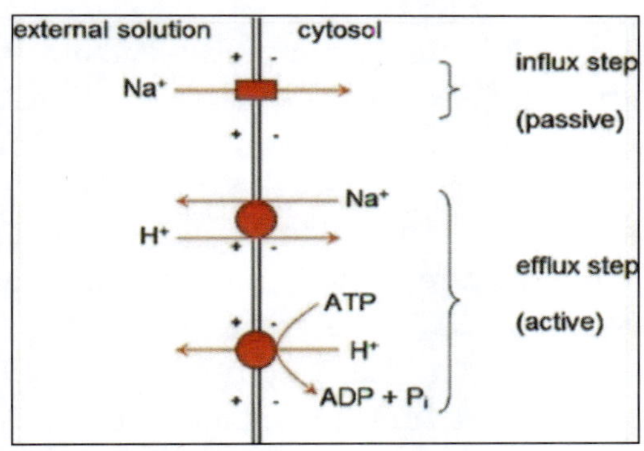

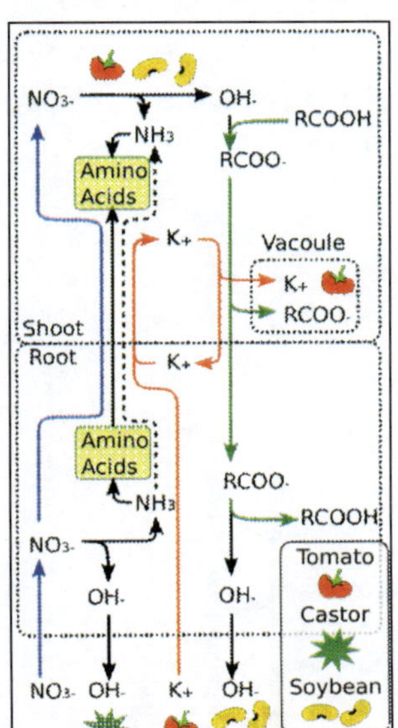

Nitrate assimilation and transport

Every nitrate ion reduced to ammonia produces one OH^- ion. To maintain a pH balance, the plant must either excrete it into the surrounding medium or neutralize it with organic acids. This results in the medium around the plants roots becoming alkaline when they take up nitrate. To maintain ionic balance, every NO_3^- taken into the root must be accompanied by either the uptake of a cation or the excretion of an anion. Plants like tomatoes take up metal ions like K^+, Na^+, Ca^{2+} and Mg^{2+} to exactly match every nitrate taken up and store these as the salts of organic acids like malate and oxalate. Other plants like the soybean balance most of their NO_3^- intake with the excretion of OH^- or HCO_3^-.

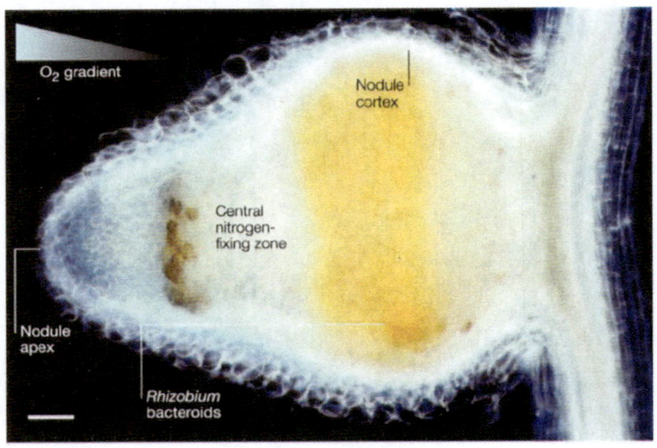

The nitrogen-fixing nodule hosts symbiotic Rhizobium bacteroids, which function as specialized nitrogen fixing organelles that exchange fixed nitrogen for photosynthates. A physiological paradox arises from the aerobic requirements of bacteroid metabolism compared with the oxygen sensitivity of nitrogenase in the absence of a dedicated bacterial protective system. Protection against oxygen is provided by the nodule environment through a cortical diffusion barrier so that the main route of oxygen diffusion is through the nodule apex, which generates a longitudinal oxygen gradient. As a result, the free oxygen concentration drops to less than 50 nM in the central nitrogen-fixing zone containing Rhizobium bacteroids. Oxygen diffusion is facilitated in the central zone by a high concentration of leghaemoglobin, and bacteroid respiration is made possible by the induction of a high affinity cbb_3 oxidase. Therefore, Rhizobium bacteroids fix nitrogen in a microaerobic, nitrogen-rich environment, which explains why oxygen, not nitrogen, is the main physiological regulatory factor for nif gene induction during symbiosis.

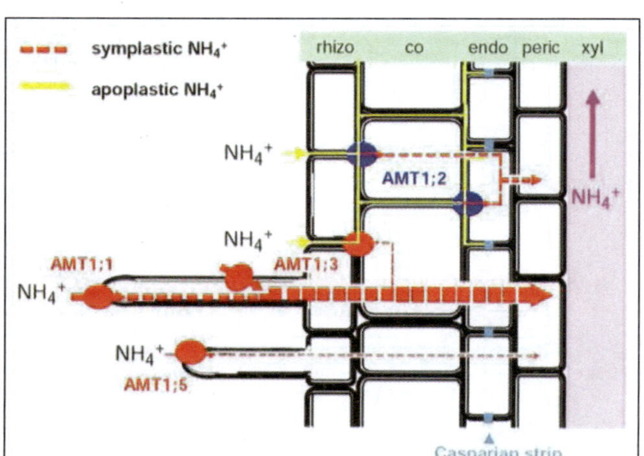

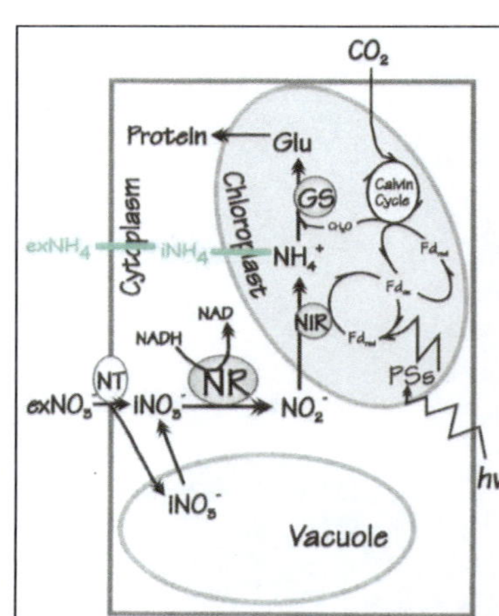

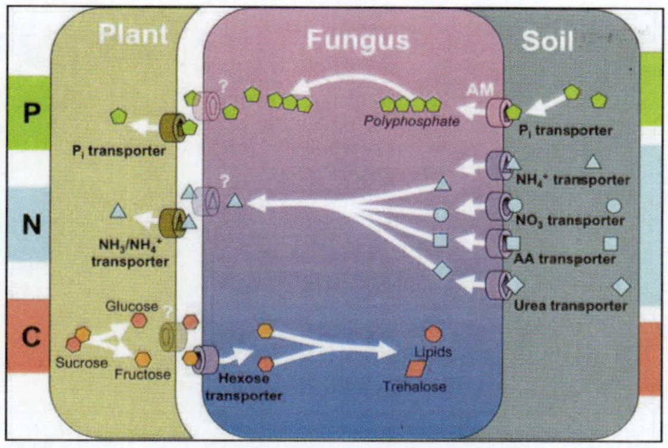

Molecular basis of nitrate uptake by the plant nitrate transporter NRT1.1

The NRT1/PTR family of proton-coupled transporters are responsible for nitrogen assimilation in eukaryotes and bacteria through the uptake of peptides. However, in most plant species members of this family have evolved to transport nitrate as well as additional secondary metabolites and hormones. In response to falling nitrate levels, NRT1.1 is phosphorylated on an intracellular threonine that switches the transporter from a low-affinity to high-affinity state. Here we present both the apo and nitrate-bound crystal structures of Arabidopsis thaliana NRT1.1, which together with in vitro binding and transport data identify a key role for His356 in nitrate binding. Our data support a model whereby phosphorylation increases structural flexibility and in turn the rate of transport. Comparison with peptide transporters further reveals how the NRT1/PTR family has evolved to recognize diverse nitrogenous ligands, while maintaining elements of a conserved coupling mechanism within this superfamily of nutrient transporters.

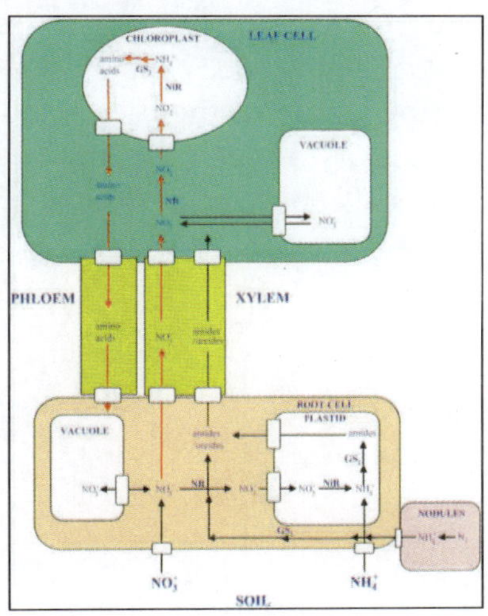

Every nitrate ion reduced to ammonia produces one OH^- ion. To maintain a pH balance, the plant must either excrete it into the surrounding medium or neutralize it with organic acids. This results in the medium around the plants roots becoming alkaline when they take up nitrate.

To maintain ionic balance, every NO_3^- taken into the root must be accompanied by either the uptake of a cation or the excretion of an anion. Plants like tomatoes take up metal ions like K^+, Na^+, Ca^{2+} and Mg^{2+} to exactly match every nitrate taken up and store these as the salts of organic acids like malate and oxalate. Other plants like the soybean balance most of their NO_3^- intake with the excretion of OH^- or HCO_3^-.

<table>
<tr>
<td>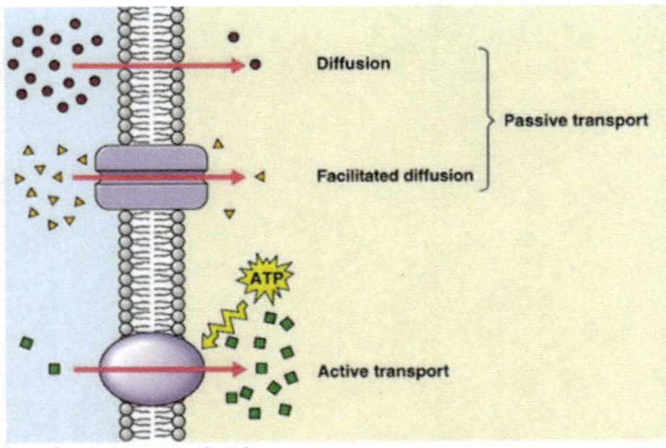
</td>
<td>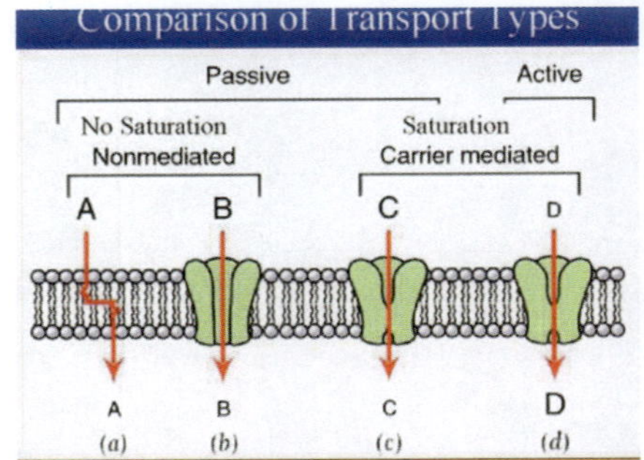
</td>
</tr>
<tr>
<td></td>
<td>Diffusion, Facilitated Diffusion, Active Transport</td>
</tr>
</table>

Membrane Permeability:

Passive transport relies on diffusion to move particles across a membrane; down their concentration gradient:

- Active transport relies on energy consumption to move particles across a membrane, up their concentration gradient.
- Transport proteins can bind to solute and move them through the membrane while selective protein channels form pores through which only specific solutes can pass.

In fact the fine micro spaces found in the cell walls never act as an impediment for the free diffusion of ions, instead the charged cellulose material greatly facilitates rapid diffusion of ion into the cell wall spaces. Thus ions diffuse through the AFS of the root system and cross through the cortical cells and outer xylem elements through the passage cells. In xylem cells, the concentration of ions is always low because of continuous upward movement of sap by transpiration pull. So the initial uptake is very rapid and a passive process. Moreover this entire event is a kind of mass movement of ions along the concentration gradient till they reach the plasma membrane.

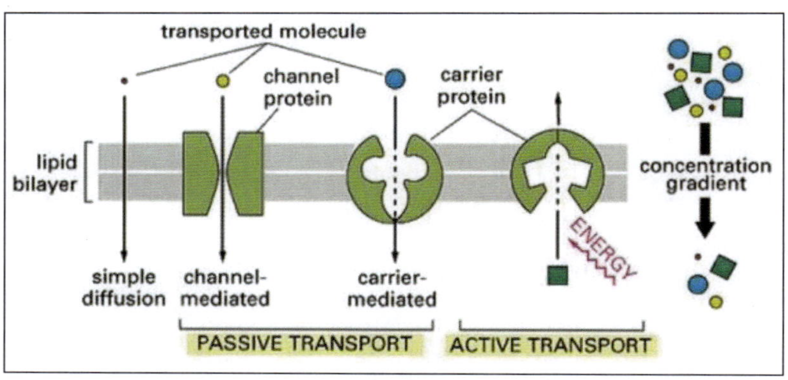

Essential cell biology

It is not just mineral uptake is facilitated by protein transporter; even water is carried by their own protein carriers called aquaporins. They have trans cellular domains, at one surface they have H_2O binding site, when H_2O binds, it changes its structure, in such a way water is released at the other side.

Recent electron microscopic studies of fractured and intact plasma membrane surfaces of plant, animal and bacterial origin reveal that the membranes contain an array of globular proteins of different sizes and dimensions scanning the entire cross-section or a part of the membranes. Some of them are aggregated in one site and others are randomly distributed. However the position of such granules in the membranes is never constant because of the fluid nature of the membrane. A large number of proteins with their 3-D globular structures act as carrier proteins or transport proteins. Such carrier proteins have high affinity towards specific ions; because they do contain specific binding site or sites.

In the past 10-15 years or so a quite a number of carrier proteins have been isolated and purified to homogeneity. The 3-D structure of a few of these proteins has been determined. Some of the common examples for carrier proteins are K, Na-ATPase pumps, hydrogen protein pumps, calcium transporting protein called Calmodulin, phosphate carrier proteins, etc.

Interestingly enough, the discovery of some cyclic polypeptides like valinomycin, nistatin, cyclosporin, etc, from many lower fungi has given a new dimension to the carrier mediated ion or nutrient uptake. Such compounds are called as 'Ionophores', because when they are added to the medium, they penetrate into the membrane easily and create a holes, which permit the entry of only one specific kind of ions. For example, valinomycin greatly facilitates the uptake of K^+ ions only. Similarly nistatin allows specific anions. It is now speculated that the plasma membranes contain such specific ion transporting complexes and they are responsible for the movement of certain ions; under certain circumstances. Ionophores role can be used as a generalized process of ion uptake, but they too play a significant role in the entry of specific ions in some species.

Using mutation as the tool molecular biologists have discovered that there are specific genes for specific carrier proteins which transport a specific substance. And such carrier proteins have been detected in the plasma membrane of the cell. Inorganic ions, monosaccharide, amino acids and many such components including proteins and nucleic acids have been found to be transported across the membrane facilitated by specific membrane proteins. Some of the receptor proteins have been found to carry the stimulants across the membranes.

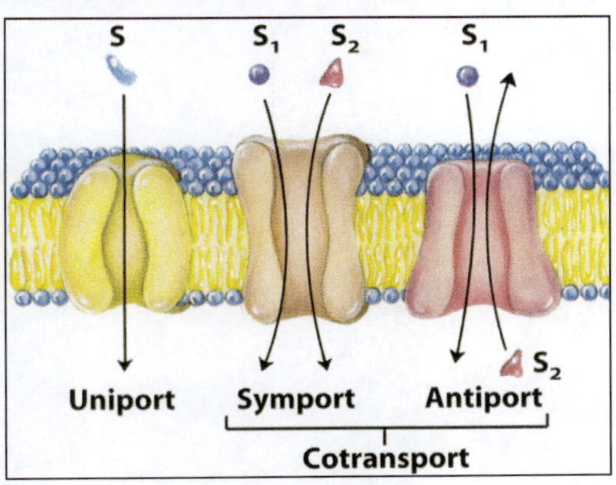

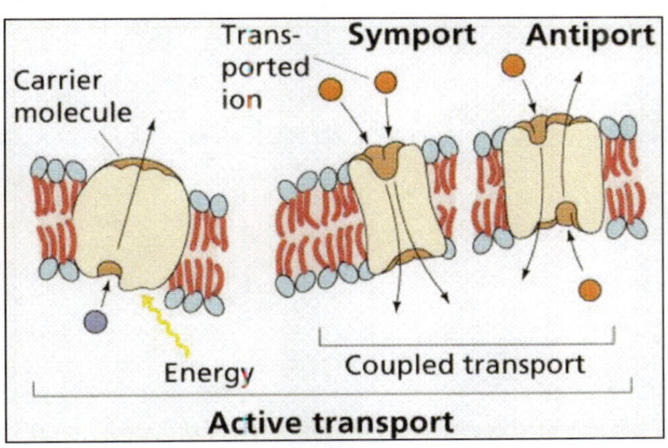

Most of the experimental evidences suggest that ion uptake is carrier mediated and the carriers are proteins. Only proteins can explain certain properties like ion antagonism, specificity of ion uptake, saturation effect, kinetic properties requirement of metabolic energy, etc. The carrier proteins exhibit 3-D shape. They have specific binding sites for the ions or substrates and no other ion can bind to such sites. After binding the proteins undergo rapid conformational change in their 3-D structure and exhibit mobility in the lipid core of the membrane. For its active transformations and transport they require metabolic energy. The activity of carrier proteins can be regulated by effectors or affecters, similar to that of allosteric enzymes. The transport activity of these proteins can be inhibited by competitive or non competitive inhibitors. Another important feature of carrier proteins is their ability to transport molecules against concentration gradient by active process.

Peter Mitchell a Nobel laureate has proposed three possible mechanisms by which carrier proteins operate in the transportation of ions or other nutrients, namely, uniport, antiport and symport mechanisms. In fact, in the same cell all the system are found and all of them operate.

Uniport Mechanisms

In this case, the activated carrier proteins pick a specific nutrient at one surface and translocate across and unload the same on the other side. After unloading, the carrier proteins revert back to its position by conformational change. Many bacterial and animal cells have ATPase dependent carrier proteins which transport many nutrients like glucose by uniport mechanism. There are many such carrier proteins which transport nutrients passively but along with the concentration gradient. Such a process is called facilitated transport but it is passive. Such uniport mechanism has been observed with respect to the transport of many monosaccharides, amino acids, etc. A very good example for this is calcium binding protein.

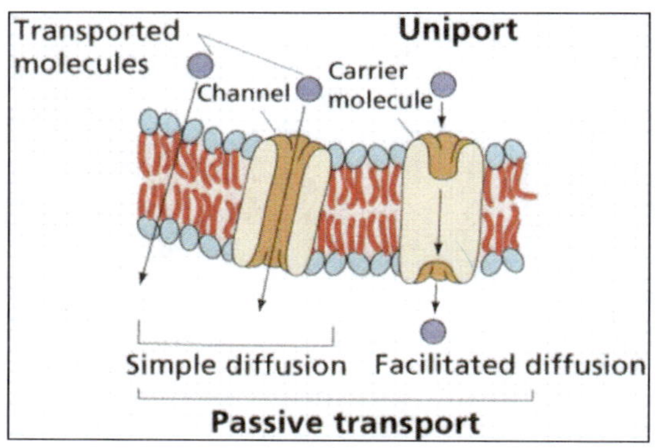

Antiport Mechanism and Symport Mechanism

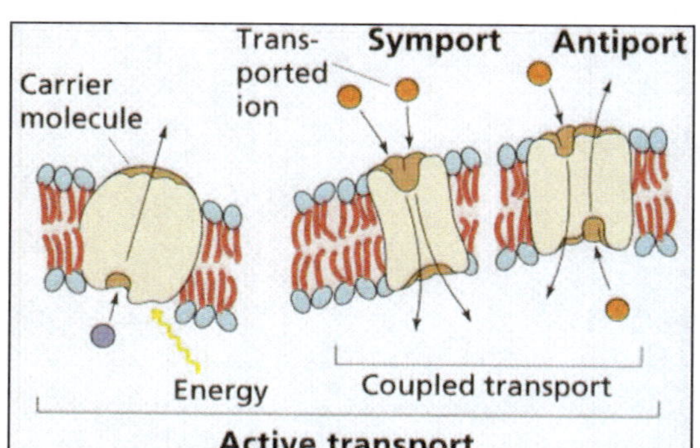

This process involves coupled transport where one specific ion is transported in and the other is transported out by the same transporter. The antiport mechanism is very well exhibited by Na/K ATPase pump found in the plant cells like Nietella and in animal cells like erythrocytes. The internal concentration of K^+ ions is many folds higher than the concentration of K^+ ions found in the extracellular fluid. Cells have to maintain such high concentrations of K^+ ions intracellularly, because they are required for various metabolic activities and for the maintenance of turgidity of the cells. On the contrary, the concentration of Na^+ ions is always higher in the external fluid than inside the cell.

Hence Na ion often leaks in or taken in by other processes. As excess sodium ions inside the cell is toxic and the same are expelled out. Na/K-ATPase pump does this by antiport mechanism.

Transport of this kind involves the binding of two different molecules or ions to two active sites of the same carrier protein. By virtue of the binding of both the ions the carrier complex rotates by conformational change. There by the said ions are released simultaneously at the opposite side of the membrane. Such symport type of transportation has been recorded in a variety of organisms. But the best example for symport mechanism is the transport of Na^+ and glucose found in animal and bacterial cells.

In this case, the carrier protein is a complex macromolecule and possesses two sites, one for Na ion and the other for glucose. While the binding sites of the protein are facing the external fluid, both Na^+ and glucose bind to their specific sites. The binding is facilitated in the sense the binding of one molecule promotes the binding of the other. However, as soon as Na and glucose bind to the protein, it undergoes conformational change and opens at the inner surface and the Na^+ and glucose are released into the cytoplasm almost at the same time. The release of molecules makes the protein revert to its original shape and position. The Na^+ ions that enter into the cell are expelled by Na/K ATPase pump, i.e. by antiport mechanism.

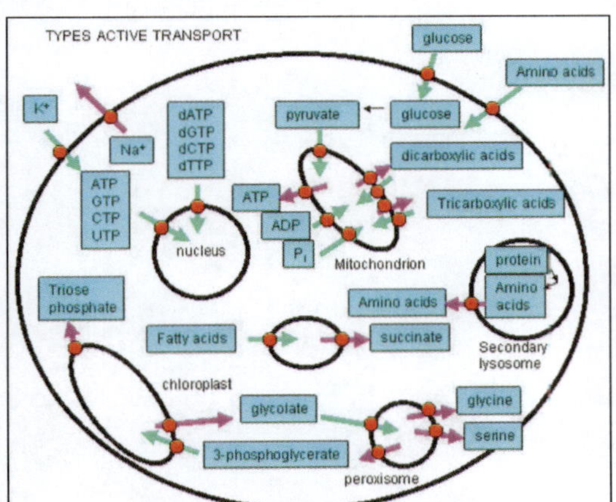

Many such species must be transported into the cell or into intracellular organelles against a concentration gradient.

Molecules must often move across membranes against a concentration gradient - from low to high chemical potential - in a process characterized by a positive DG. As protons could be "pumped" across the inner mitochondrial membrane against a concentration gradient, powered by the DG associated with electron transport (passing electrons from NADH to dioxygen), other species can cross membranes against a concentration gradient - a process called active transport - if coupled to ATP hydrolysis or the collapse of another gradient. This active transport is differentiated from facilitated diffusion we studied earlier, which occurred down a concentration gradient across the membrane. Many such species must be transported into the cell or into intracellular organelles against a concentration gradient.

Pinocytotic Transport Mechanism

Pinocytosis is not just restricted to animal cells only, it is also found in plants and perform the functions similar to that of animal cells. This process is an energy dependent process and involves the uptake of nutrients in bulk.

Plasma membrane on its outer surface possesses binding sites for ions or nutrients in the form of clusters in specific areas. Throughout the surface one finds such groups of binding sites. Each of these clusters may contain only one kind of binding sites or different kinds. Once the ions bind to their specific sties, the inner surface of the plasma membrane in such regions, coated pits gets activated. Involvement of Clathrin coated and the activity of microfilaments that are associated on the face pull the membrane inside by contractile activity. Thus the inwardly puckered membrane with all the solutes and some water is drawn inwards and ultimately pinched off as solute loaded vesicles. Later these vesicles are transported intracellularly or they may break down into smaller vesicles. The contractile activity which is responsible of Pinocytosis is an energy dependent process.

Translocation of Mineral Salts

Absorbed mineral salts and other components like cytokinins, and others absorbed by roots, ultimately reach the xylem elements found in the root system i.e xylem vessels. Most of the minerals are absorbed by meristems than root hairs. The minerals then are transported to the vascular system of young and developing xylem elements or they can be transported into mature vascular elements.

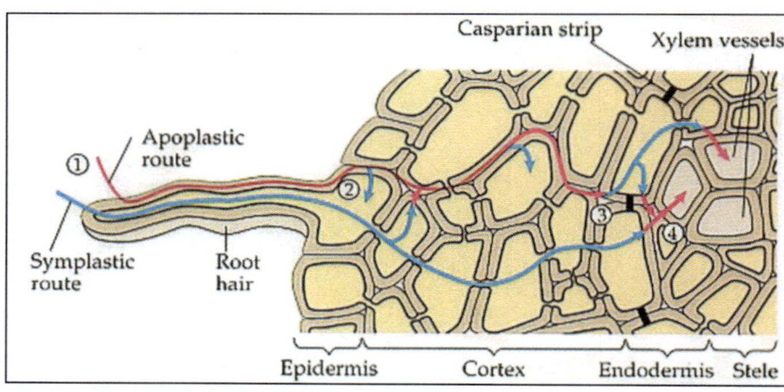

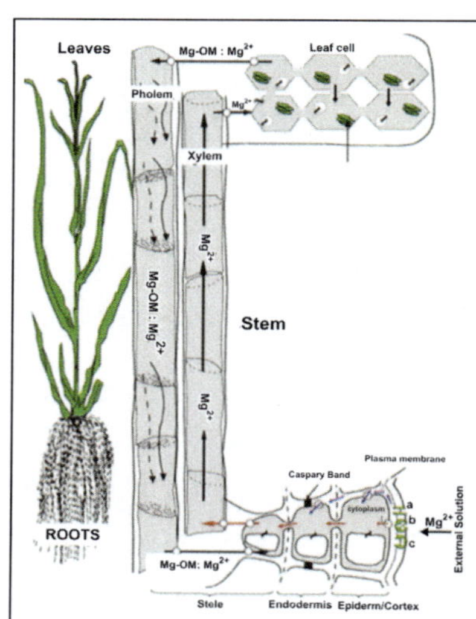

The xylem elements have xylem parenchyma, which are living and supporting the process of translocation of minerals into upward conducting system. Once the elements are loaded into xylem cells, they are transported upwards along with the transpiration stream and thus they reach the aerial parts of the plant body where the nutrients; first diffuse into apparent free spaces then they are absorbed into living cells by active/facilitated transport. While the minerals are translocated all along the length of the stem, considerable amount of them diffuse laterally and reach the cortical cells.

In spite of the process of absorption of mineral salts by root system is highly specific and energy dependent, the upward movement is entirely due to passive process for the dissolved components just move along with the water column due to transpiration pull. Yet one cannot rule out the ability of livings cells to do such active transportation for dead xylem are supported by living xylem parenchyma. Even Xylem parenchyma tissue a living tissue has roles in such mineral uptake and transportation. Whichever factor that affects the transpiration pull also affects the translocation of mineral salts in the plant body.

References

- Soil-types, geology: lumenlearning.com, Retrieved 18 July, 2019
- The-building-blocks-of-soil, soil-basics, basic-agronomy: vaderstad.com, Retrieved 20 August, 2019
- Soil-structure, soil-basics, basic-agronomy: vaderstad.com, Retrieved 22 january, 2019
- Structure-forming-processes, soil-basics, basic-agronomy: vaderstad.com, Retrieved 24 February, 2019
- Reading-soil-types, chapter, geology: lumenlearning.com, Retrieved 26 March, 2019
- Characteristics-of-different-soil-types, soil-basics, basic-agronomy: vaderstad.com, Retrieved 28 April, 2019
- Soil-acidity-definition-sources-and-problems, soil-acidity: soilmanagementindia.com, Retrieved 11 January, 2019
- Plant-Cellular-Physiology, Absorption-Of-Mineral-Nutrients: plantcellbiology.masters.grkraj.org, Retrieved 30 May, 2019

Chapter 2

Understanding Soil Fertility

The ability of soil to sustain agricultural plant growth is known as soil fertility. It results in sustained and consistent high quality yields along with providing plant habitat. To maintain the soil fertility, it is very important to maintain the quality of irrigation water. This chapter discusses in detail the theories and methodologies related to soil fertility.

Soil fertility may be defined as the inherent capacity of soil to supply plant nutrients in adequate amount and suitable proportion and free from toxic substances. Soil productivity is the ability of the soil to produce crop per unit area. Thus a fertile soil may or may not be productive depending upon crops, marketing condition and several other factors (excessive acidity or alkalinity, the presence of toxic substances, poor physical properties or an excess or deficiency of water. But every productive soil has to be fertile. Soil productivity mainly depends on soil fertility to a great extent.

The soil is a natural medium for plant growth and it supplies nutrients to plants. Some soils are productive and they support luxuriant growth of plants with very little human effort whereas others may be unproductive which support almost no useful plant life regardless of every human effort In order for soil to be productive, it must:

1. Be easily tillable and fertile.
2. Contain all essential elements in the forms readily available to plants in sufficient amount.
3. Physically good to support plants and contain just the right amount of water and air for proper root growth. The soil must supply these essentials every day in the life of the plant.

A soil can be highly fertile, i.e., it has ready supply of nutrients in available form, yet it may not be highly productive. Water-logged soils may be highly fertile but may not produce good crop because of the un-favourable physical conditions.

A fertile soil may be highly saline or alkaline which may not be good for agriculture Sandy soil may be poor in fertility but with the use of fertilizers and water it may be made productive. Soil fertility thus denotes the status of plant nutrients in the soil whereas the soil productivity is the resultant of various factors influencing crop production. In fact there is no standard for either fertility or productivity because both depend upon the crops to be grown. Soil that is productive for potatoes may not necessarily be productive for certain other crops.

Plant Nutrients

Of the 90 or so chemical elements forming the earth's crust, 16 are known to be essential for plant growth and reproduction. Seven elements needed in good quantity (macro nutrients) are hydrogen, oxygen, nitrogen and carbon from air and water and phosphorus, potassium and calcium from mineral particles in the soil. The other 9 elements needed only in small amount (micronutrients) are magnesium, sulphur, boron, copper, iron, manganese, zinc, molybdenum and chlorine.

Cobalt and sodium are essential elements for plant growth. With the exception of hydrogen, carbon, and oxygen all other inorganic plant requirements are obtained directly or indirectly from the soil minerals, hence these elements are called mineral nutrients. Strictly speaking, nitrogen is not a mineral element but it has been included in the list because it can also be obtained by plants from soil.

The mineral elements are taken by plants from soils mostly in the form of ions. Plants obtain nutrients from the following four devices:

1. From the soil solutions through roots,
2. From exchangeable ions on the surface of clay and humus particles through roots,
3. From readily decomposable minerals, and
4. Through the leaves.

The essentiality of an element is proved by the following criteria:

1. The element may be considered essential if its exclusion from the nutrient medium inhibits or drastically reduces the growth and reproduction of plant.
2. Acute deficiency of the element produces certain well defined disease symptoms which are not produced by the deficiency of any other element.
3. Deficiency disease symptoms will disappear if the particular element is supplied before the living system has been damaged beyond repairs.

The capacity of soil to supply the essential elements is a fundamental edaphological problem, In natural habitat the plant matter is returned to ground through decay and unless there is extensive leaching and percolation the inorganic nutrients become available again for new growth. Many farm crops are removed from ground in to and the supply of essential elements thereby becomes depleted.

To replace these elements taken away from ground by crops, the manures or fertilizers are added. Unless it is done the crop will suffer from deficiencies in the essential elements. Deficiencies become reflected in the growth of plants in several ways; some may cause a reduction in yield as a result of poor plant growth and some may delay maturation of crop a function that may be very vital to crop yield in the places where the growing season is short.

The symptoms of mineral deficiencies may be dwarf, spotted, distorted, curled or wilted leaves or rotting of the centre of fruits. Different mineral nutrients have certain specific deficiency symptoms.

Soil Fertility Evaluation

Soil fertility evaluation is the process of estimating the amount of native and residual nutrient elements which could be available for use by growing crops in particular soil and the amount of fertilizer to be supplemented for profitable crop production. Soil fertility evaluation, therefore is a tool for.

- Determining the fertilizer needs of specific crops and soils;
- Achieving reliable and economic fertilizer recommendations, that is, ensuring that right types and quantities of fertilizers are applied;
- Checking wastage of fertilizers; and
- Minimizing soil and water pollution through the addition of excessive amounts of chemical fertilizers. Although soil fertility evaluation is a powerful tool to support high productivity by way of rationalizing nutrient use, its current impact on farm practice is presently not visible. In order to make it an effective and farmer oriented service, it is imperative to:
    - Expand the arena of soil fertility evaluation beyond NPK, and as well as the pH of the soil.
    - Develop fertilizer recommendations for high yield targets, involving all deficient nutrients and exploiting important positive nutrient interactions.

Methods of Evaluating Soil Fertility

There are three basic tools for evaluating soil fertility. They are listed below based on their relevancies, starting with the least useful.

- Visual symptoms of nutrient deficiency.
- Plant tissue analysis.
- Soil analysis.

Using visual symptoms of nutrient deficiency to determine fertilizer needs:

Visual nutrient deficiency symptoms can be a very powerful diagnostic tool for evaluating the nutrient status of plants. One should keep in mind, however, that a given individual visual symptom is seldom sufficient to make a definitive diagnosis of a plant's nutrient status. Wade argued that many of the classic deficiency symptoms such as tip burn, chlorosis and necrosis are characteristically associated with more than one mineral deficiency and also with other stresses that by themselves are not diagnostic for any specific nutrient stress. However, their detection is extremely useful in making an evaluation of nutrient status.

In the vast majority of cases, nutrient deficiencies can substantially reduce production *without showing any clear symptoms*. This problem is referred to as "*hidden hunger*" whereby a deficiency is having a negative effect without being recognised, though if an early diagnosis is made, effective action can usually be taken. The principal advantages of visual diagnostic symptoms are that they are readily obtained; can provide an immediate indication of nutrient status and where the symptoms do not require confirmation no apparatus of any kind is necessary. Its main drawbacks are that the visual symptoms do not develop until after there has been a major effect on yield, growth and development. Symptoms may be complicated or suppressed by other factors, such as salinity and non-uniformity of nutrients, weather conditions and pests or disease organisms and the complications may lead to forming an entirely wrong conclusion. Nevertheless, with experience, an observant farmer can learn to use the visual method quickly and with great advantage. Such farmer must be very familiar with the basic theoretical knowledge of nutrients deficiencies.

Plant Tissue Analysis

Plant tissue analysis is a laboratory determination of the total elemental content of plants or of certain plant parts. It is used for a variety of purposes including monitoring the nutrient status of crops an troubleshooting problem areas. It also serves as the basis for nutrient recommendations for perennial fruit crops). It is the only way to know whether or not a crop is adequately nourished during the growing season. Plant tissue analysis should not be confused with tissue testing. Tissue testing typically refers to a field test that involves taking sap samples from fresh plant tissue and analyzing the samples on site. Plant tissue analysis is performed on dried plant tissue that has been processed in a laboratory Plant tissue analysis can detect unseen deficiencies, confirm visual symptoms of deficiencies and detect toxic levels of nutrients. Though usually used as a diagnostic tool for future correction of nutrient problems, plants tissue analysis from young plants will allow a corrective fertilizer application that same season. The most important use of plant analysis is to monitor nutrient status and diagnose existing nutrient problems. as well as to keep an excellent yearly record of crop nutrient use and needs under different environmental conditions. However, it is a tool which must be used with caution. The tissue sampling method is critical for success. The procedure is unique to each crop. The plant must be at specific stage of growth, and specific tissue must be selected.

Nutrient Element

Deficiency Symptoms in Crops

- Nitrogen: General chlorosis (appearance of light-green to pale-yellow colour starting from the older leaves). The tips are first affected. Older leaves scorch or drop depending on the degree of deficiency. Growth becomes stunted and plant may have etiolated appearance. In acute case, flowering is greatly reduced. Deficient crops have lower protein content.

- Phosphorus: The mature leaves have characteristic dark-green to blue-green coloration. The root development is restricted, plant is spindly and stunted. In acute case, red, purple or brown leaf coloration develops and maturity is delayed. There is poor development of seed and fruit.

- Potassium: Slow and stunted growth of plants. Chlorosis along the leaf margins followed by scorching and browning of tips of older leaves. These symptoms gradually progress inward. Stalks are weak and plants lodge easily. Terminal and lateral buds may die ("dieback"). Shriveled seeds of fruits.

- Calcium: These are first noticed in the meristematic regions and the young leaves. The young leaves of new plants are often distorted, small and abnormal. Leaves may be cup-shaped and crinkled and the terminal buds deteriorate with some breakdown of petioles. Root growth is severely impaired. Roots get rotten. Buds and blossoms shed prematurely. Stem structure weakened.
- Magnesium: Inter-veinal or marginal chlorosis, mainly of older leaves often accompanied by development of a variety of pigments. Chlorosis may also begin in patches or pouches which later merge and spread to the leaf margins and tips. With acute deficiency, the affected tissue may dry up and die. Leaves are usually small, brittle in final stages and curve upwards at margin. Twigs weak and prone to fungus attack, usually premature leaf drop.
- Sulfur: Younger leaves turn uniformly yellowish or chlorotic. Plants are spindly and grow poorly. Flower production is often indeterminate. Stems are stiff, woody and small in diameter.
- Zinc: Deficiency symptoms mostly appear on the 2nd or 3rd fully mature leaves from the top of plants. In some species, leaves may be dark-green or blue-green. Flowering and fruiting are much reduced under conditions of severe zinc deficiency, and the entire plant may be stunted and misshapen.
- Boron: Growing tips are often damaged and may die. The leaves may become distorted, curling and becoming brittle. Stems become rough and cracked; often with corky ridges of spots. Flowers do not form. Roots are stunted and prone to bacterial infection. Brown heart in root crops characterized by dark spots on thickest part of the root or splitting at center. Fruits such as apples develop" internal and external cork" symptoms.
- Copper: Leaves may become chlorotic (yellowing) or deep blue-green. Curling of leave blades or rolling up of margins. Flowering and fruiting are curtailed. Ear production in cereals restricted and grains poorly set. Young shoots often die back, whereupon new shoots emerge from multiple buds further back, making for a bushy appearance. Annual plants may fail to develop and may die at the seedling stage.

Some visual systems of selected nutrient deficiencies in cereals (maize).
Zinc defiency; Magnessium, sulphur and iron; N, P and K defiencies

Some visual systems of selected nutrient defiencies in tomato

Failure to follow the prescribed method for that crop will produce misleading results. The analysis may be of little importance if plants come from field that are infested with weeds, insects, disease organisms; if the plants are stressed for moisture; or if plants have some mechanical injury. As posited by Steinhilber and Salak and other earlier researchers that there are rules that must be followed for an analysis to be successful, a simple mistake can be deleterious for the crops concerned.

Once the analysis is done, the nutrient contents are compared with known minimum values for that crop (critical values) or sufficiency ranges and nutrient deficiencies or excesses are identified. The critical values and sufficiency ranges for several crops are shown in table below. The lower limit (of sufficiency range) represents the critical level below which appropriate fertilizer rates should be applied. The upper level indicates the level above which toxicity sets in. Figure is a graphical depiction of interpretive guidance for a generic crop. The nutrition concentration ranges (% or ppm) for each nutrient status category and the shape of the growth curve differ from crop to crop.

Table: Sufficiency range of nutrient elements in some crops.

Source/amount of fertilizer (bags)							
S no.	Crop	Nitrogen (N)	Phosphorus (P)	Calcium (Ca)	Magnesium (Mg)	Zinc (Zn) (ppm)	Boron(B) (ppm)
1	Maize (whole plants, less than 30 cm tall)	3.5-5.0	2.50-4.0	0.30-0.70	0.15-0.45	20-60	5-25
2	Maize (leaf below the whorl prior to teaseling. Ear leaf at teaseling collected before silks turn brown)	Leaf below the whorl:3.0-3.50	Leaf below the whorl:2.0-0.50 Ear Leaf: 1.75-2.25	0.25-0.50	0.13-0.30	15-50	4-25

3	Cotton (upper mature leaves on vegetative stems taken prior to bloom or when first squares appear).	3.50-4.50	1.50-3.0	2.0-3.0	0.30-0.90	20-200	20-60
4	Groundnut (upper mature leaves taken prior to or at boom stage)	3.50-4.50	1.70-3.00	1.25-2.00	0.30-0.80	20-60	20-60

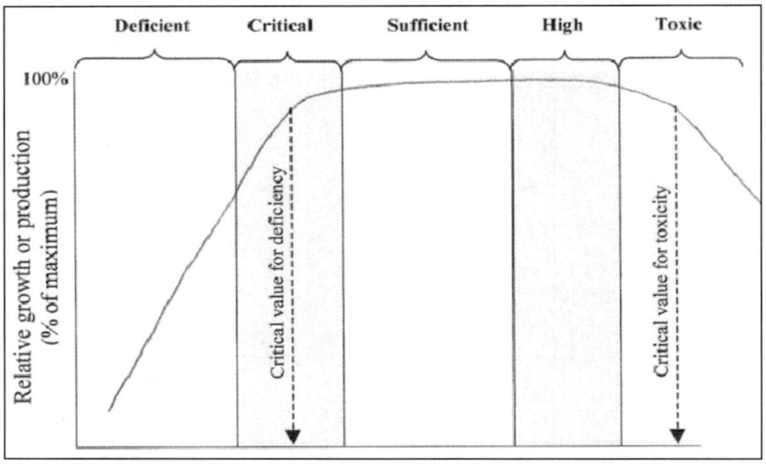

Australasian Soil and Plant Analysis Council Inc. Modified in Plant Analysis: an Interpretation Manual

Although very detailed, plant tissue analysis results cannot be used as a sole determinant for generating nutrient recommendation for crops. It is therefore pertinent to look at the primary source of the nutrient – the soil, rather than the tissue. The result of the soil test will be very important in determining nutrient needs of crops.

Soil Testing/Soil Analysis

Soil testing is used to determine both the amount of each nutrient that is immediately available and the amount that can become available during the life of a crop. Various methods have been developed and the key to success is that the methods must be calibrated. Soil test calibration implies establishing relationship between soil test values and relative crop response. Soil sampling done properly forms the basis of a successful long-term soil and crop nutrient managemen plan. It is most useful before planting to predict lime or fertilizer needs. Also, it measures levels of specific nutrients in a soil. However, it cannot indicate whether plants growing in that soil are able to take up the nutrients. Soil test are the best way to assess soil pH.

Objectives of Soil Testing

1. To accurately determine the status of available nutrients in soils (P, K, Mg, pH, Zn, B).
2. To clearly indicate to the farmer the seriousness of any deficiency or excess that may exist in terms of various crops.
3. To form the basis on which fertilizer needs are determined.

4. To express the results in such a way that they permit an economic evaluation of the suggested fertilizer recommendation.

How to Sample Soils?

It is not possible to move the entire soil of the farm to the laboratory for analysis. Only a small sample is required. A good sample is the first requirement for a reliable soil test. This sample should be a true representative of the farm/plot/field, that is, it should contain all the characteristics of the soil on this farm/plot/field. The proper methods of collecting and handling samples are determined by certain factors:

1. Accuracy and precision.
2. Sample areas that are representative of the farm.
3. Effect of farm size on accuracy.
4. When, how deep and how often to sample.
5. The use to be made of the analyses.
6. The pattern and ease of recognition of soil variability.
7. Previous and proposed management practices.

Some soil sampling tools

Although many types of sampling designs exist only two main types (random and systematic) are commonly used in the soil and earth sciences. Simple Random gives opportunity to all samples to be involved in the final selection while in stratified simple sampling; points are assigned to predefined groups or strata and a simple random sample chosen from each stratum. Stratified sampling (correctly applied) is likely to give a better result than simple random sampling.

Soil sampling could be done before or after preparing the land for planting (that is, to estimate pre-plant fertilizer needs). The farm operator must decide what level of detail is relevant to his or her field operations. Are there parts of the field that have different fertility patterns? Are these areas large enough to be relevant? Does the operator want to engage in site- specific management? Has the operator the ability to vary fertilizer application rates to accommodate the field subsections identified? The farm should be divided into small units of 1 hectare in size. Each unit should have uniform observable properties. Subsections of a field would commonly be identified by differences in topography (termed landscape -directed soil sampling), parent material, management history, or yield history. It may be impossible to subdivide a field into smaller units if the farm operator has no prior knowledge of the field, or if there is no obvious topographic or parent material differences. A composite sample consisting of 10 to 20 spot (core) samples collected using approved sampling techniques is obtained for each unit of the farm.

Tools

The best tool is the metal tube called "sampling tube". If this is not available, cutlass, shovel, hand trowel and auger could be used. Do not use brass, bronze, or galvanized tools because they will contaminate samples with copper and zinc.

Soil test kit: This is a compact soil testing equipment with full complement of devices and reagents for the determination of the pH, electrical, conductivity, nitrogen, phosphorus and potassium in soil, fertilizer and water. Carbon, which is a good index of nitrogen content in soil, can also be determined. The equipment is robust and cost saving in terms of laboratory space.

Method of Soil Testing

To sample a farm: A rough map of the farm dividing it into sampling units as shown in figures was made. A composite soil sample is taken from each soilsampling unit. The farm was then sampled as indicated in the following illustrations:

1. Use the right sampling tool: The best tool is a metal tube called a sampling-tube. However, if this is not available, any of the materials illustrated below could be used. These include cutlass, shovel, hand trowel and augur.
2. The sample: A composite sample; comprising of 10 to 20 core (single location) (However 15 sub-samples may be the minimum number required to give sufficiently low variance, especially for P) samples that have been taken randomly in a zigzag fashion across the same soil area (land unit) is obtained. The land unit or soil area from which each composite sample is obtained is about one hectare. Sampling fertilizer bands, terrace channels, dead furrows, roads and other unusual areas should be avoided.

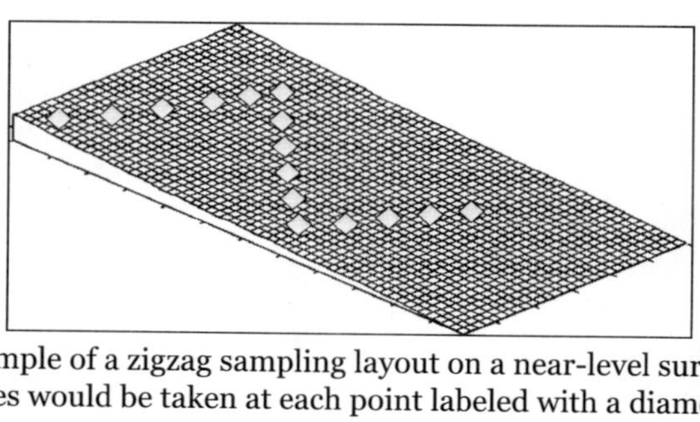

Example of a zigzag sampling layout on a near-level surface
Soil samples would be taken at each point labeled with a diamond shape.

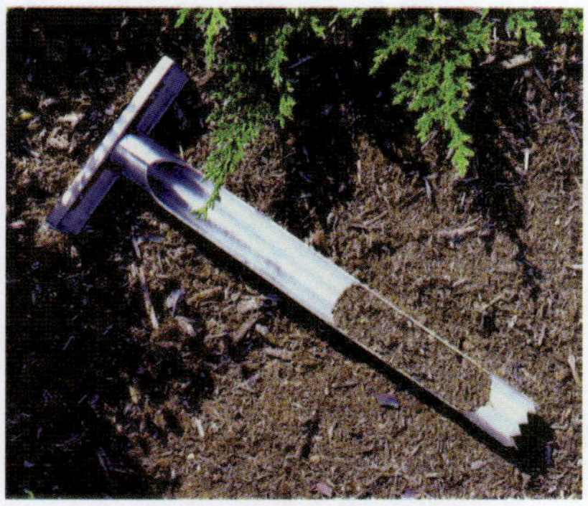

Collecting soil sample using sampling tube

Collecting soil samples using a shovel, spade or cutlass

Mixing of soil samples after collection

Getting the sample ready for laboratory test

(a) Sample the soil from the surface to about 15 cm (6 inches) depth with a sampling tube, or any of the tools stated above. For a sampling tube, the process is as illustrated in figure.

(b) If the tool available is a spade, a shovel, or a cutlass, proceeds as follows:

- Dig a v-shaped hole, 15 cm (6 inches) deep.
- hen take one-and a –quarter centimeter (1/2 inch) slice of soil sample from the smooth side of the v-shaped hole illustrated in figure.

3. Put all the core samples taken from the soil area together in a clean plastic bucket, as a composite sample.
4. Mix the sample well with a clean rod or with your hand in the bucket.
5. Pour the soil sample into a clean plastic bag and tie it securely.
6. Label each plastic bag of sample properly.
7. Fill out the information requested as accurately as possible and send the sample to laboratory for analysis.

Soil Test Result Interpretation

The figures on the soil analysis report do not indicate the exact amount of nutrients available to a crop but when interpreted correctly give a description of the soil fertility status. The analytical result is used to suggest how much nutrient should be applied. The exact amount needed will depend on the crop to be grown and must be modified to suit the conditions under which it is grown.

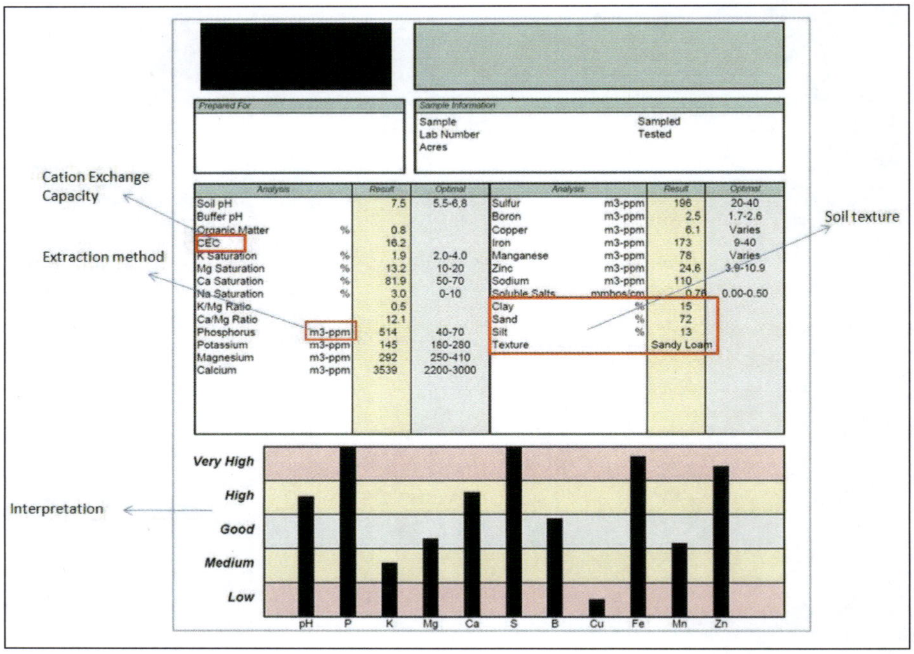

Sample of soil test result

Continuous removal of nutrients from the soil via different means requires continuous replacement to maintain productivity. This replacement (fertilization) requires specific knowledge in order to truly maximize yield, minimize cost and to reduce adverse effect on soil crops. Of the methods available, soil test seems to be the easiest to predict fertilizer requirement for Nigerian farmers. At a certain level, the complex tissue analysis may be used as a tool but it must be combined with soil test result.

Soil Fertility Management

Advancing food security and environmental sustainability in farming systems requires an integrated soil fertility management approach that maximizes crop production while minimizing the mining of soil nutrient reserves and the degradation of the physical and chemical properties of soil that can lead to land degradation, including soil erosion. Such soil fertility management practices include the use of fertilizers, organic inputs, crop rotation with legumes and the use of improved germplasm, combined with the knowledge on how to adapt these practices to local conditions.

Soil Fertility

Soil fertility is the most important asset of a nation. Maintenance of soil fertility is an important aspect of agriculture. The soil fertility problem has been studied in many countries and scientists have brought to light several facts concerning soil fertility and its maintenance.

Approaches to Efficiently Manage Soil Fertility

An integrated soil fertility management aims at maximizing the efficiency of the agronomic use of nutrients and improving crop productivity. This can be achieved through the use of grain legumes, which enhance soil fertility through biological nitrogen fixation, and the application of chemical fertilizers.

Whether grown as pulses for grain, as green manure, as pastures or as the tree components of agro-forestry systems, a key value of leguminous crops lies in their ability to fix atmospheric nitrogen, which helps reduce the use of commercial nitrogen fertilizer and enhances soil fertility. Nitrogen-fixing legumes are the basis for sustainable farming systems that incorporate integrated nutrient management. Use of nitrogen-15 lends understanding of the dynamics and interactions between various pools in agricultural systems, including nitrogen fixation by legumes and utilization of soil and fertilizer nitrogen by crops, both in sole and mixed cropping systems.

Soil fertility can be further improved by incorporating cover crops that add organic matter to the soil, which leads to improved soil structure and promotes a healthy, fertile soil; by using green manure or growing legumes to fix nitrogen from the air through the process of biological nitrogen fixation; by micro-dose fertilizer applications, to replenish losses through plant uptake and other processes; and by minimizing losses through leaching below the crop rooting zone by improved water and nutrient application.

The Contribution of Nuclear and Isotopic Techniques

The isotopes of nitrogen-15 and phosphorous-32 are used to trace the movements of labelled nitrogen and phosphorous fertilizers in soils, crops and water, providing quantitative data on the efficiency of use, movement, residual effects and transformation of these fertilizers. Such information is valuable in the design of improved fertilizer application strategies. The nitrogen-15 isotopic technique is also used to quantify the amount of nitrogen fixed from the atmosphere through biological nitrogen fixation by leguminous crops.

The carbon-13 isotope signature helps quantify crop residue incorporation for soil stabilization and fertility enhancement. This technique can also assess the effects of conservation measures, such as crop residue incorporation on soil moisture and soil quality. This information allows the identification of the origin and relative contribution of different types of crops to soil organic matter.

Types of Soil Fertility

Inherent or Natural Fertility

The soil, as a nature of them, contains some nutrients which is known as 'inherent fertility'. Among the plant nutrients, nitrogen, phosphorus and potassium is essential for the normal growth and yield of crop. Indian soil contains 0.3 to 0.2 per cent nitrogen, 0.03 to 0.3 per cent phosphorus and 0.4 to 0.5 per cent potassium. The inherent fertility has a limiting factor from which the fertility is not decreased.

Acquired Fertility

The fertility developed by application of manures and fertilizers, tillage, irrigation etc. is known as 'acquired fertility'. The acquired fertility has also a limiting factor. It is found by experiment that the yield does not increase remarkably by application of additional quantity of fertilizers. So it is necessary to apply fertilizer on the basis of nutrient content of a soil and it is estimated by soil testing.

Losses of Plant Nutrients from the Soil

Soil is the store house of plant nutrients. Losses of plant nutrients from the soil are the main cause of decreasing the fertility of soil.

Plant nutrients are lost from the soil by the following ways:

- Removal of Plant Nutrients by Harvested Crops: Plant absorbs nutrients from the soil and stores them in their different parts. The crops remove large quantity of nitrogen and potassium and relatively small quantity of phosphorus. Four to five per cent of total nitrogen is lost from soil per acre annually through the harvested crops. The loss can be reduced by adding farm waste materials to the soil.
- Removal of Plant Nutrients by Weeds: Weeds compete with crops for mineral nutrients. Competition begins when the supply of plant nutrients falls below the requirements of both weeds and crops. Weeds by nature of them grow fast and remove the plant nutrients from the soil. This will be acute if weeding is not done as soon as weeds emerge or germinate.
- Losses of Nutrient by Soil Erosion: Erosion is the physical removal of top soil by water and wind. Plant nutrient, particularly nitrogen remains on the upper layer of soil. When erosion is severe, the nutrient is lost along with soil and the fertility of soil decreases accordingly.
- Losses of Nutrients by Leaching: Fertilizers, both straight and mixed, are soluble in water and as such they are liable to loss by leaching in rain water or irrigation water. Leaching loss is more acute in sandy soil and bare soils. Nitrogen is mainly lost from soil by leaching.
- Losses of Nutrients in Gaseous Form: Nitrogen is generally subjected to loss in gaseous form. Losses of nitrogen occur due to following reduction:
    - Denitrification: The biochemical reduction of nitrate nitrogen to gaseous compounds is called 'denitrification' and is thought to be the most widespread type of volatilization. The reactions are believed to occur as follows:

Nitrate Assimilation

$$NH_2\,OH \rightarrow NH_3\ \text{(Ammonia)}$$

$$\uparrow$$

$$\underset{\text{Nitrate}}{NO_3} \rightarrow \underset{\text{Nitrite}}{NO_2} \rightarrow (HNO) \rightarrow H_2N_2O_2 \rightarrow N_2O\ \text{(Nitros oxide)}$$

$$\text{or} \qquad \downarrow$$

$$NO_2\,NH_2 \rightarrow N_2\ \text{(Elemental Nitrogen)}$$

Denitrifiction

All these changes apparently are directly biochemical and are encouraged by poor aeration and drainage and the presence of abundant amounts of easily affected nitrogenous compound. The biochemical reduction of nitrate (NO_3) and nitrite (NO_2) to gaseous nitrogen is often termed denitrification.

- Nitrites in slightly acid solution will evolve gaseous nitrogen when brought in contact with certain ammonium salts with simple amines such as urea and even with non-nitrogenous sulphur compounds and carbohydrates. The following reaction is suggestive of what may happen to urea:

$$\underset{\text{Nitrous acid}}{2HNO_2} + \underset{\textit{Urea}}{CO(NH_2)_2} \rightarrow \underset{\text{Carbon dioxide}}{CO_2} + \underset{\text{watar}}{3H_2O} + \underset{\text{Elemental nitrogen}}{2N_2}$$

This type of gaseous loss is strictly chemical and does not require either the presence of microorganism or adverse soil condition.

Pseudomonas fluorescence, P. denitrificans, P. stutzeri, Bacillus subtitis, Thiobacillus denitrification etc. are associated with denitrification.

Factors affecting Soil Fertility

The factors affecting soil fertility may be of two types; i.e. – (a) Natural factor and (b) Artificial factor. The natural factors are those which influence the soil formation and the artificial factors is related to the proper use of land.

The factors affecting the fertility of soil are as follows:

i. Parent Materials: The property of soil depends on the property of parent rock. If the parent rock contains more nutrient, the soil developed from rock contains more nutrient. The soil developed from calcareous rock contains more phosphorus than the soil which is developed from granite rock. The soil developed from acid igneous rock (i.e. quartz) and basic igneous rock (i.e. norite and dolerite etc.) become sandy and clayey in nature respectively.

ii. Climate and Vegetation: Plant kingdom is closely related with climate. Rainfall and temperature has an effect on soil fertility. In heavy rainfall areas, the nutrients are lost by leaching. As a result of which the fertility of that soil becomes low. Besides these, the upper layer is eroded which decreases the soil fertility. Organic matter is oxidised in high temperature. For this, the fertility of soil in temperate region becomes low.

iii. Topography: The fertility of soil is also dependent on the topography of soil. Leaching and erosion is most common in sloppy land. As a result of which, the fertility of that soil becomes low. On the other hand, the fertility of level and becomes more, because the nutrient of high land in soluble form deposited in the level land, specially in low land.

iv. Inherent Capacity of Soil to Supply Plant Nutrients: The nutrient contents of a soil vary according to the nature of soil. The soil which contains much quantity of nutrient becomes more fertile. In an experiment in central farm, Coimbatore, it was found that the garden soil of nine inches depth contains 1400 lb (630 kg) potassium per acre. So the fertility of soil depends on the inherent capacity of soil.

v. Physical Condition of Soil: Aeration and movement of water is good in the soil containing adequate amount of organic matter and this type of suitable condition of soil is beneficial for the growth of plant. The physical condition of soil should be suitable for the growth and development of plant. It is essential for proper supply of oxygen in the soil. Improper supply of oxygen is unsuitable for the growth of plants as well as for the proper function of soil organism.

As a result of which, organic matter do not decompose properly and the nutrients of organic matter does not transformed in available form of the plant. Suitable physical condition of the soil increases the water holding capacity of soil which is favourable for the growth of plant. The fertility of soil depends mostly on the texture and structure of soil.

vi. Soil Age: The soil developed earlier losses its fertility gradually. Because the fertility of the soil decreases by the process of leaching and weathering in course of time. Besides this, cultivation of crops without manuring decreases the fertility of soil.

vii. Micro-Organism and Soil Fertility: Various types of organism live in the soil. The soil organism brings the unavailable nutrients into the available form. Different types of bacteria, fungi and algae live in the soil. The nitrifying bacteria fix nitrogen from air. Dr. P.K. De, in an experiment showed that blue green algae fix 50 kg nitrogen per hectare in the paddy land having good amount of water.

viii. Availability of Plant Nutrients: The nutrient of the soil must be in the available form of plant. The plant do not absorb nutrient if it is not soluble in water. Super phosphate applied in acid soil is converted into iron or aluminium phosphate which is not soluble in water. As a result, phosphate remains in the soil in unavailable form to plant.

The availability of nutrients depend mainly on the following factors:

a. Soil aeration.

b. Soil pH.

c. Activity of microorganism.

ix. Soil Composition and Fertility: The plant absorbs the nutrient from the soil. The nutrient of minerals becomes available by weathering. The soil containing more organic matter becomes more fertile. The sandy soil is less fertile, whereas loamy soil is more fertile.

x. Organic Matter and Soil Fertility: The fertility of soil increases if the soil contains more organic matter. Organic matter contains the plant nutrients. Besides this, organic matter improves the physical condition of soil. Decomposition of organic matter increases the nitrogen content of soil. Thus the fertility of soil increases.

xi. Soil Erosion: Erosion is the physical removal of top soil by water and wind. As such it decreases the fertility of soil. Because the nutrients remaining in upper layer of soil is lost by erosion and the fertility of soil decreases accordingly.

xii. Cropping System: Cultivation of same crop year after year in the same field decreases the fertility of soil. There are various types of cropping system in India such as mono-cropping, mixed cropping, relay cropping and crop rotation. Crop rotation increases the fertility of soil.

xiii. Favourable Environment for Root Growth: The suitable condition for growth of plant depends on physical, chemical and biological condition of soil. Soil contains 25 per cent water and 25 per cent air by its volume and this condition is favourable for good aeration. The bad aeration in the soil is not good for the growth of the crop.

Importance of Soil Fertility

Most of the time, the average person treats the soil "like dirt". A wise farmer/rancher will care for the soil because he knows that man is dependent on the top 6 inches (15.2 centimeters) of soil. In the plant-animal-soil continuum, soil is often neglected because it does not indicate stress in an obvious way. Animals and plants show physical symptoms but the soil must be looked at more carefully to monitor good health.

Soil that is rich in nutrients is fertile. The expectation of growing plants as food for livestock must include the reality that plants will take nutrients out of the soil. Replacing nutrients is the basic goal of fertilization. Improper fertilization in the past has caused controversy, but the basic premise of fertilization is to replenish the soil.

Soils feed the plants which in turn feed the animals that feed us. Including soil in this important chain will help guarantee its success. Soil provides the support or foundation for plants and most of the nutrients. Soil is accumulated decomposing plant and animal matter with aging parent material. As the soil components break down, elements are released and become available to plants as nutrients. However, naturally this process takes a long time and the soil will only be a result of the parent material, climate, those living organisms once living there, topography, and time. So what is made available to a plant at a certain time may not be exactly what a growing plant needs. Fertilization is supplementing the existing soil with additional, needed nutrients. Fertilizing wisely increases yield, quality (nitrogen content and digestibility), and profits.

There are three basic ways to replenish the nutrients removed from the soil. One way is to recycle nutrients, mainly by way of animal waste. This is a crucial method when discussing pastures. Another way to replenish soils is to obtain and apply fertilizer. And the third way is through microbial action such as nitrogen fixation. Forage-livestock producers should understand and utilize all three for maximum fertility with minimum cost and environmental damage.

Fertilization is an important issue because it is needed in order to produce enough food for the increasing population from the decreasing cultivated land, but too much or inappropriate use can be detrimental to the environment.

Effect of Irrigation on Soil Fertility

Water is a very important natural resource, which is the basis of all life forms. For sustainable agricultural production, water is one of the most precious important inputs. Plants need water in huge amount throughout their life. Water is also one of the main factors that influence most of the metabolic process such as photosynthesis, respiration, adsorption, opening and closing of stomata and translocation of food material. The growth and yield of crop plants is very much affected by the availability of water.

Water is used in Two Ways

- Withdrawal or off-channel use.
- Non-withdrawal or on-channel use.

The amount of water taken out of a streams or pumped out from underground water reservoir or surface water reservoirs to be supplied to the points of major use, such as public water supply systems, irrigation and industries, is referred to as off channel use. In non-withdrawal use, water is used without being removed from its natural source such as for navigation, swimming bodies, wildlife habitats and other recreation purposes.

Need for Irrigation

Irrigation is the artificial watering of soil to sustain plant growth. Irrigation has become necessary because of the limitation of using natural rainfall as the reliable source of water for agriculture. It is also difficult to store the rainwater for immediate irrigation purposes.

Through the monsoon season usually lasts for four months, most of the rains occur only during two months. Therefore, it is very difficult to plan the cropping pattern depending solely on the rams. The sufficient recharging of ground water is essential for sustainable farming, but usually this objective is not fulfilled due to unpredictable and fluctuating rains.

The efficient utilization of water resources is essential for better crop production. It includes the suitability of land and water for irrigation, planning of crops and suitable water management practices. Water management includes irrigation and drainage.

The suitable irrigation depends on:

- Time of irrigation.
- Amount of irrigation.
- Efficient method of irrigation.

Similarly, suitable drainage depends on:

- How much to drain.
- How best to drain.
- How rapidly to drain.

Stages of Crop when Irrigation is Required

The growth span of a crop plant passes through various phases and stages of growth. The rate of irrigation varies during the different stages of plant growth, i.e., from seedling to maturity stage. The growth period of irrigated crops can generally be divided into three phases, namely vegetative, flowering and maturity stage. At vegetative stage, light and intermittent irrigation is required whereas at flowering stage moderate and frequent irrigation is needed and during crop maturation stage again light irrigation is required.

Systems of Irrigation

There are different water sources from which irrigation is carried out, for example, canals, wells, open wells, tube wells, tanks etc.

Depending upon the crops, soil types, water resources, climate conditions and costs involved, several systems of irrigation are used which are as follows:

- Surface irrigation systems.
- Subsoil irrigation systems.
- Drip irrigation systems.

Surface Irrigation System

In this system, water is directly used on the surface of the soil and water follows the slope of the land, the surface irrigation system may be of following types:

Flood Irrigation

Flood irrigation is used for close-grown crops such as rice and where farm fields are leveled and water is abundant. A sheet of water is allowed to advance from different sources and remains on a field for a given period, depending on the crop, the porosity of the soil and its drainage.

Different surface irrigation systems Flood irrigation (b) furrow irrigation (c) basin irrigation

Basin Flooding

It is used in widely spaced trees, such as in orchards, with basins built around trees and filled with water.

Check Basin

This system of irrigation is widely used in India, as it is more suitable to all types of soils and to a variety of crops.

Furrow Irrigation

In furrow irrigation, water moves in the fields in furrows, between two ridges. It is employed in the fields with row crops such as cotton and vegetables. Parallel furrows, called corrugations, are used to spread water over fields that are too irregular to flood.

Sprinkler Irrigation System

In this system, irrigation is done through pressure, to the surface of any crop or soil, in the form of a thin spray. Sprinkler irrigation system uses less water and provides better control. Each sprinkler, spaced along a pipe, sprays water in a continuous circle until the moisture reaches the root level of the crop. Centre-pivot irrigation uses long lines of sprinklers that revolve in a circular field. It is used especially for feed crops such as alfalfa, tall crops and orchards.

1. Sprinkler irrigation system conveys water from the source to the field, through pipes under pressure, and distributes over the field in the form of spray of 'rain like' droplets.
2. 10-20% of area is not irrigated at the comers of square of rectangular plot.
3. This system requires high energy and involves huge cost of the equipment.

This system can be followed in conditions when:

1. The soil is too shallow.
2. The land is too steep.
3. Less and frequent irrigations are needed.
4. The soils are very sandy.

This system is disadvantageous in the following conditions when:

1. Strong winds cause improper distribution of water.
2. Evaporation losses are high from sprinkler irrigation, especially under high temperature and low relative humidity conditions.
3. The initial cost is high.

Sprinkler irrigation: a. center–pivot irrigation with long line of sprinkler b. sprinkler irrigation system in which water is conveyed to field through pipes under pressure

Sub-Soil Irrigation System

In this system, irrigation is done into a series of ditches in the field deeply to the impervious layer. Then it moves laterally and vertically through capillaries saturating the root/one. In this system, continuous supply of water in the root zone is assured from the artificial water table created by the ponding of irrigation water on the impervious layer. This system is very efficient because the water losses through evaporation from surface can be reduced. This system is more common in Gujarat and Jammu and Kashmir, for cash crops growing on sandy loam soils.

Drip Irrigation System

It is also called trickle irrigation. This irrigation system involves the slow application of water, drop by drop, to the root zone of a crop. It is done through mechanical devices called emitters, located at selected points along water delivery lines. Drip irrigation is extensively used in areas of acute water scarcity and especially for crops such as coconut, grape, banana, citrus, sugarcane, cotton, maize, tomato and plantation crops.

Drip irrigation

Advantages of Drip Irrigation

1. Water loss through transpiration is low.
2. It is possible to obtain better yield and quality of crops by controlling soil moisture-air-nutrient levels.
3. We can save the fertilizers by monitoring the supply of nutrients as per the need of the crop.
4. Improvements in biological fertility can be achieved by avoiding pollution.

Disadvantages of Conventional Irrigation Methods

1. Major part of the water goes waste, only small quantity of water is utilized by the plants.
2. Water is not uniformly distributed due to uneven and poor leveling of the field.
3. Crops are usually subjected to cyclic changes of flooding and water stress situations by providing heavy irrigation at one time and leaving the fields to dry up for about 10 to 15 days.

4. The low lying fields always get excess water that causes prolonged water logging, due to lack of leveling of fields.
5. In the fields, about 10-15% of land is utilized for making channels and distributaries etc. which decreases the area of cultivation.
6. Excessive irrigation and poor water management leads to water logging and accumulation of salty in the upper layer of the soil.
7. In conventional irrigation, unmanageable and undesirable weeds grow.

Problems of Excess Irrigation

1. Excess irrigation causes several changes in the soil and plants, resulting in reduced growth and sometimes even death of plants.
2. Germinating seeds are sensitive to water logging, since they are totally dependent on the surrounding soil space for oxygen supply.
3. Yield of cereals is reduced if excess irrigation is given.
4. Excess water causes injury to the plant due to low oxygen supply to the root system and accumulation of toxic substances in the soil.
5. Leaching of nitrates and de-nitrification occur which result in nitrogen deficiency.
6. Permeability of roots decrease due to shortage of oxygen. It results in decreased uptake of water and nutrient.

References

- Soil-fertility-definition-types-and-factors-soil-science, soil-fertility: soilmanagementindia.com, Retrieved 1 June, 2019
- Soil-fertility-its-meaning-causes-and-maintenance, soil: biologydiscussion.com, Retrieved 3 July, 2019
- Soil-fertility-evaluation, potential-tool-for-predictingfertilizer, requirement-for-crops: researchgate.net, Retrieved 5 August, 2019
- Improving-soil-fertility: iaea.org, Retrieved 7 January, 2019
- Soil-fertility-definition-types-and-factors-soil-science, soil-fertility: soilmanagementindia.com, Retrieved 19 Feb February, 2019
- Fertility, fertilization: oregonstate.edu, Retrieved 21 March, 2019

Chapter 3

Types of Fertilizers

Fertilizer is a material that is applied to soils or to plant tissues to supply one or more nutrients which are integral to the growth of the plant. Fertilizers can be natural or synthetic. The topics elaborated in this chapter will help in gaining a better perspective about the various types of fertilizers.

Chemical Fertilizers

Of the elements known to be essential for plant growth, nitrogen (N), phosphorus (P) and potassium (K) are required by plants in pretty large amounts, and are therefore, designated as major ox primary nutrients while calcium, magnesium and sulphur are secondary nutrients. For acid soils, use of Ca and Mg is necessary. Seven elements iron, manganese, boron, molybdenum, copper, zinc and chlorine are required in trace amount and hence called micro-nutrients.

Under continuous cultivation our soils are losing organic matter and mineral nutrients faster than they can be replaced. Regular loss of nutrients from the soil results in compact soil, shallow roots, increased drought, daddy and poorly productive soil. So, for the maintenance of soil fertility quick replacement of the organic matter and mineral nutrients removed from the soils is necessary.

Nitrogen (N), phosphorus (P) and potassium (K), i.e., the primary plant nutrients are commonly applied to soils in the form of commercial fertilizers and hence they are often referred to as fertilizer elements.

Table: Fertilizers and their composition.

	Fertilizer	Composition	source	Nitrogen%
1	Ammonium sulphate	$(NH_4)SO_4$	By –produce from coke and also as synthetic	26.6
2	Ammonium nitrate	NH_4NO_3	synthetic	33
3	Ammonium chloride	NH_4CI	synthetic	25
4	Ammonium liquor	Dilute NH_4OH	synthetic	20-25
5	Ammonium phosphate	$NH_4H_2PO_4$	synthetic	11% N and 48% P_2O_5
6	Diammonium phosphate (DAP)	$(NH_4)_2HPO$	synthetic	21% N and 53% P_2O_5
7	Sodium nitrate	$NaNO_3$	Chile saltpetre and synthetic	16
8	Calcium ammonium nitrate	NH_4NO_3 and Dolomine	synthetic	20.5

9	Calcium Cyanamid	$CaCN_2$	synthetic	21
10	Calcium nitrate	$Ca(CN_3)_2$	synthetic	15
11	Urea	$CO(NH_2)_2$	synthetic	45

Chemical fertilizers are classified into the following three group on the basis of materials supplied:

- Nitrogenous fertilizers.
- Phosphorus fertilizers.
- Potassium fertilizers.

Classification is not so simple as this grouping would imply, several such fertilizer materials as contain two of these elements, as for example, potassium nitrate and ammonium phosphate.

Nitrogenous Fertilizers

Crops usually take nitrogen from the soil in the form of nitrate (NO_3^-)and ammonium ions (NO_4^+).

Nitrogen fertilizers may be divided for convenience into two groups:

- Organic nitrogen fertilizers.
- Inorganic nitrogen fertilizers.

Organic Nitrogen Fertilizers

Organic materials such as cotton seed meal, guano and fish tank age are nitrogen carriers but because they supply less than 2% of total nitrogen added in commercial fertilizers, their use is costly and hence they are not used extensively. Nitrogen of organic fertilizers is released slowly by microbial action. They are used as special fertilizers for gardens, lawns and potted plants.

Inorganic Nitrogen Fertilizers

Several inorganic chemicals are used to supply nitrogen to plants.

Phosphorus Fertilizers

Phosphorus has rightly been called 'master key' to agriculture as low crop production is due more often to lack of phosphate than to the deficiency of any other element except nitrogen. In phosphorus fertilizers this element is present in the form of phosphate or superphosphate salts and it is available to the plants when it is combined with organic matter or with calcium and magnesium. Phosphorus is also found in combination with iron and aluminium and is present in certain rock minerals as apatite.

Plants take up phosphorus chiefly as phosphate $(PO_4)^-$, HPO_4^- and $H_2PO_4^-$ ions and the availability of these ions depends chiefly upon the acidity of ground. They become nearly insoluble in strongly acid or strongly alkaline soils. Release of phosphorus from phosphate rocks is slow. The breakdown of phosphate fertilizers produces phosphoric acid (P_2O_5) in soluble form that is absorbed by plants.

Use of phosphate fertilizers on alkaline soils is not suitable. Phosphorus fertilizers are classified into (i) water soluble, (ii) citrate soluble, and (iii) insoluble. When the term P.O, is used it means water soluble plus citrate soluble (P_2O_5). The following are the important phosphate fertilizers being used in all parts of the world including tropical Asia.

Table: phosphate fertilizers.

Fertilizer with chemical composition	P_2O_5 Available percentage
(a) Fertilizers containing water soluble phosphorus	
(i) Superphosphate $Ca(H_2PO_4)_2$ and (ordinary grade) $CaHPO_4$	16-20
(ii) Superphosphate $Ca(H_2PO_4)_2$ and $CaHPO_4$ (concentrated)	40-45
(iii) Ammonium phosphate $NH_4H_2PO_4$	48 (11%N))
(iv) Diammonium phosphate $(NH_4)_2HPO_4$	46-53 (4%N)
(b) Citrate soluble phosphorus fertilizers	
(i) Dicalciam phosphate	35-40
(ii) Basic slag (indian) $(CaO)_5P_2O_{5-}SiO_2$	3-5
(c) Insoluble phosphorus fertilizer	
(i) Rock phosphate fluoride and chloraphatite	20-30
(ii) Bone meal $Ca_3(PO_4)_2$	18-20

Superphosphate

It is water soluble fertilizer. It does not affect the soil adversely. It contains mono-calcium phosphate, di-calcium phosphate and tri-calcium phosphate, gypsum, silica, iron aluminium sulphate and calcium fluoride and water.

Ammonium Phosphate

It is a fertilizer containing both nitrogen and phosphorus. It is rich in phosphoric acid content but comparatively low in nitrogen content. Ammonium superphosphate ($NH_4H_2PO_4$). Ca_3 $(PO_4)_2$ $(NH_4)_2$ SO_4. It is the cheapest fertilizer and is a mixture of nitrogen and phosphorus fertilizers. It contains nitrogen 3 to 4 per cent and phosphorus pentaoxide (P_2O_5) 16 to 18%.

Nitro-phosphate

It is highly hygroscopic. It contains nitrogen 13-18 per cent and phosphorus 20 per cent. It is suitable for acid soils.

Bone Meal

It is derived from bone. Bone ash and bone char are the bone products. It is suitable for acidic soils.

Basic Slag

Basic slag, a by-product in the manufacture of steel, is one of the cheapest sources of phosphorus. It is a double compound of silicate and phosphate of lime. It is dark brown in colour and alkaline in reaction.

Rock Phosphate

It occurs in natural deposits. It is light grey in colour. It is a very cheap fertilizer suitable for acid soils.

Potassium Fertilizers

These fertilizers are soluble in wate which means that potassium is readily available to plants. Total potassium of potassium fertilizers is usually expressed in terms of water soluble potassium (K) or potash (K_2O) Soils of arid and semiarid areas are generally well supplied with potassium. Acid soils usually need potassium fertilizers more than neutral or alkaline soils because acid soils develop in the areas of high rainfall that leaches out available potassium. All potassium fertilizers are physiologically neutral in reaction.

The following are some common potassium fertilizers:

- Potassium chloride: It is also called muriate of potash. It contains 48—62% K_2O. It is also cheap and neutral in reaction.
- Potassium sulphate: It contains about 50% potash (K_2O). It is expensive fertilizer.
- Kainite: It is natural potassium mineral which contains 14—20% potassium. It is suitable for alkaline soils.
- Wood ashes: It is used in the form of ash as a manure. Potassium occurs in the form of potassium carbonate and the percentage of potash is from 2 to 6.5. It is suitable for alkaline soils.

Application of Micronutrients

In order to correct the deficiency of micronutrients, especially if it is very necessary, micronutrients should be added only after ascertaining the amount required. Copper, manganese, iron zinc are supplied generally as their sulphate and boron is applied as borax. Molybdenum is supplied as sodium molybdate. In recent years there has been an increase in use of chelates to supply iron, zinc, manganese and copper.

Soils vary in their ability to supply available nutrients. Some soils may be deficient in nitrogen, some may be deficient in nitrogen and phosphorus, and still some others may be deficient in nitrogen, phosphorus and potassium. To suit the variable requirements of different soils and crops, fertilizer mixtures are prepared. Fertilizer mixtures or mixed fertilizers contain two or more fertilizer materials. If the ingredients and their amounts are known, the formula is referred to as open and if they are not disclosed the formula is termed closed one.

The kinds and amounts of fertilizers to be applied to soil are determined considering the following points:

- Kinds of crop to be grown—particularly its economic value, nutrient removal and absorbing ability.
- Chemical condition of soil in respect of total nutrients and available nutrients.
- Physical state of the soil, especially its moisture content and aeration. For recommending the kind and amount of fertilizers or soil amendments, the analysis of soil is essential.

Application of Soil Conservation Practices

Loss of plant nutrients and water from the soils due to soil erosion can be checked effectively and the fertility of soil can be maintained by application of various biological and engineering methods of soil conservation.

Water Management Practices (Irrigation and Drainages)

Water supply is critical factor in crop production in most areas of the world. Soil moisture greatly affects the availability of mineral nutrients in the soil. It has been proved beyond doubt that fertilizer response is much higher with adequate irrigation.

Drainage and moisture control influence micronutrient availability in soils. Improving the damage of acid soils encourages the formation of less toxic oxidized forms of iron and manganese.

Prevention and Elimination of Inorganic Chemical Contamination of Soil

Loss of soil fertility due to application of toxic chemicals as pesticides can be eliminated if:

- Application of toxic chemicals to soil is reduced.
- The soil and crop are so managed as to prevent cycling of toxic chemicals.

Stabilization of Soil pH

The stabilization of pH through application of soil amendments and buffering seems to be an effective guard against the problems of non-availability of certain plant nutrients and radical changes in microbial activities arising due to change in soil pH.

Uses of Chemical Fertilizers

Chemical fertilizers are synthetic substances that provide the necessary nutrients for healthy plant growth and development. Chemical fertilizers combine synthetic forms of the primary macronutrients for plants such as nitrogen, phosphorus and potassium, as well as a variety of micronutrients and additives. Synthetically developed fertilizers provide the benefit of having more nutrients per pound than organic fertilizers, thus reducing costs. Chemical fertilizers have many uses for the average gardener and homeowner, as well as for commercial use.

The nutrients in chemical fertilizers help plants achieve lush, green growth.

Gardening

Plant and vegetable gardeners use chemical-based fertilizers for healthy plant growth. The nitrogen in fertilizer promotes cell division and leafy vegetation, phosphorus helps plants with photosynthesis and respiration and potassium helps plants absorb the other nutrients, according to NASA. Chemical fertilizers contain a higher concentration of macro- and micronutrients than organic fertilizers, which allows gardeners to spread less of a concentration for the same results. Depending on brand, chemical-based fertilizers provide varying levels of additives, which help grow plants faster and stronger. However, according to Pennsylvania State University, over-fertilization, a common problem with chemical-based fertilizers, can cause poor growth and even plant death. University or private labs can test soil samples to determine the recommended amount of fertilizer to apply for specific garden plants.

Lawn Care

The nutrients synthetically added to chemical fertilizer help with lawn care by promoting quick, lush green growth. However, over-fertilization can cause tip burn and brown patch areas on grass. According to the University of Minnesota, inorganic nitrogen-producing compounds found in chemical fertilizer such as ammonium nitrate, calcium nitrate and ammonium sulfate are likely to contribute to lawn burn. The chemicals, released with the application of water, create quick results but may cause damage. Fertilizer for lawns, as well as the gardening varieties, comes in two main forms, quick release and slow release. Both varieties have their advantages and disadvantages for plant growth.

Quick Release

Quick-release chemical fertilizers, made from compounds that are water-soluble, allow plants to absorb nutrients immediately after application. This process produces noticeable plant growth over a short period. However, according to the University of Minnesota, quick-release fertilizers are more likely than slow-release fertilizers to cause tip or plant burn. Due to the fast absorption, quick release fertilizers are applied on a systemic basis more often than slow-release fertilizers are.

Slow Release

Slow-release fertilizers release nutrients over an extended period. According to the University of Arizona, slow-release fertilizers release soluble nitrogen by either a hydrolysis or microbial process. This application allows for gardens, lawns or crops to have continual nutrients without regular fertilization. Slow-release fertilizers are less likely to cause plant or lawn burn but are easily washed out by heavy rainfall.

Organic Fertilizers

Unlike commercial chemical fertilizers, organic fertilizer for gardens is typically made of single ingredients and can be matched to your garden's particular nutritional needs. The types of organic fertilizer can come from plant, animal or mineral sources, depending on what chemicals your garden requires. In order to qualify as an organic fertilizer, the materials must naturally occur in nature. Fertilizer for organic gardening isn't the quick and instant fix that chemical fertilizers can be. With organics, you have to let moisture and beneficial organisms break down the content of the fertilizer material in order for the plants to get to the nutrients inside. In general, half of the nutrients in an organic fertilizer ingredient can be used the first year it is applied, and the rest of it is slowly released in the years to come, feeding and conditioning the soil.

Benefits of Organic Fertilizers

Organic fertilizers are carbon-based compounds that increase the productivity and growth quality of plants. They have various benefits over chemical fertilizers, which include the following:

Non-toxic Food

Use of these organic fertilizers ensures that the food items produced are free of harmful chemicals. As a result, the end consumers who eat these organic products are less prone to diseases such as cancer, strokes, and skin disorders, as compared to those who consume food items produced using chemical fertilizers.

On-farm Production

The majority of organic fertilizers can be prepared locally or on the farm itself. Hence, the cost of these fertilizers is much lower than the cost of chemical fertilizers.

Low Capital Investment

In addition to the on-farm production possibilities of organic fertilizers, organic fertilizers help in maintaining the soil structure and increasing its nutrient-holding capacity. Therefore, a farmer who has practiced organic farming for many years will require far less fertilizer, because his soil is already rich in essential nutrients.

Fertile Soil

Organic fertilizers ensure that the farms remain fertile for hundreds of years. Land located at the site of ancient civilizations, such as India and China, are still fertile, even though agriculture has been practiced there for thousands of years. Fertility is maintained because organic fertilizers were always used in the past. However, with the increased use of chemical fertilizers today, the land is rapidly becoming infertile, forcing many farmers to further increase their use of chemical fertilizers or even leave the farming industry entirely.

Safe Environment

Organic fertilizers are easily bio-degradable and do not cause environmental pollution. On the other hand, chemical fertilizers contaminate both the land and water, which is a major cause of diseases for human beings and is the force behind the extinction of a number of plants, animal, and insect species.

Employment

We all know that chemical fertilizers are made in large plants that are automated and have an annual capacity of millions of tons. Organic fertilizers, on the other hand, are prepared locally and on a much smaller scale. As a result, the production of organic fertilizers leads to employment, especially in rural areas where employment opportunities can sometimes be bleak.

Applying Fertilizers

Farmers have the conception that adding large quantities of fertilizers in their farms will only be beneficial to their plants. However, according to The Royal Horticultural Society, since fertilizers are substances that are rich in nutrients, they are meant to be added in small quantities. The society suggests the following points for fertilizer application:

- Soil Analysis: A thorough analysis of the soil should be conducted before planting the crops. You should first find out the existing nutrient content of your soil, find out the nutrient requirement of the plants you want to grow, and add fertilizers based on the nutrient deficiency between the two. A typical soil analysis test will determine the soil texture, organic matter content, and pH. It will also give you the content of different minerals such as phosphorus, magnesium, and potassium. Some soil test labs also determine the micro-organic activity in your soil.

- Time of Application: The time when you apply the fertilizers is also crucial. In cold climates, fertilizers should not be applied during winter and autumn, as they will favor young growth during the winter. Since the weather is harsh during this time, the plants won't be able to survive for long. Therefore, fertilizers should only be applied during the spring. In fact, the time for adding the fertilizers and preparation of the crop should be quite close.

- The quantity of Fertilizers: Society suggests that if a farmer is in doubt about the quantity of fertilizer to be added, he should add less fertilizer to be on the safe side.
- Organic Fertilizers: It is also suggested that farmers prefer organic fertilizers to chemical fertilizers. Global awareness for the hazards of long-term chemical fertilizer use is growing. Because of this, more and more farmers all over the world are shifting to organic fertilizers. The agricultural market has also recognized this trend and has recently employed a full-blown campaign to promote organic and natural fertilizers. Among the benefits of using organic fertilizers are non-toxic food, lower cost, better soil fertility, and of course, a safer environment.

Compost

One of the best features of organic gardening is the ability to locally produce the fertilizers, pesticides, and other organic ingredients. And the easiest among them is compost. You can prepare compost almost anywhere, even in your kitchen garden.

Agricultural scientists all over the world agree that compost is an excellent source of organic matter for garden plants. Compost provides air, water, organic matter, and microorganisms to your plants, thus enhancing their growth. It also maintains a healthy atmosphere for the soil and hence keeps insects, plant diseases, and weeds away.

Benefits of Compost

So what is so special about compost? Why not use organic matter directly? The biggest advantage of using compost is that the organic matter in compost is partially decayed, so its volume is much lower. Furthermore, the microorganism activity has started in the compost already, so the concentration of these microorganisms is very high. This makes the compost a concentrated and easy to absorb the source of organic matter for the growing plants.

Where can you buy compost? You can purchase compost from the market, as well as online stores. Compost for Sale is a directory of some companies that sell compost in the UK, US, and India.

You can also prepare compost in your own garden. Composting on a small scale can be done through the following three techniques:

Fast Composting: Fast composting is a composting technique used by many biodynamic farmers. A pit of 1m height, 1m width, and 1m length is prepared. The length of the pit can be increased according to the space available in your garden and your compost requirement. The pit is filled in with a thin layer of leaves, manure, and straws. Water is added after adding each layer. You can also add some soil between these layers. The pile should be turned regularly to keep it aerated. Compost will be ready within six to eight weeks.

Slow Composting: Slow composting, as the name suggests, is a very slow process and it takes a number of months for compost formation through this method. Organic matter that is rich in carbon (brown organic matter such as sawdust) is the main ingredient. Fill half of the composting pit with carbon-rich organic matter, and every day, add kitchen waste or vegetable peelings to this mixture. You can also add soil along with the kitchen waste occasionally. The pit's contents will gradually decompose and you can start harvesting the compost from the bottom of the pit.

Worm Composting: Worm composting, also known as vermiculture, can be carried out in a bin or in a trench. Fill the bin or trench with soil and organic matter (kitchen waste or vegetable peelings). Then, release red worms into the bin. Ensure that the bin is always moist, but don't add too much water. Adding water in large quantities will fill the pores built by the worms with water and disturb their colony. You can also make a hole in the bin for purging out any excess water. Some researchers also suggest using earthworms instead of red worms, since earthworms are more effective in converting organic matter into manure. The compost will be ready in a few weeks. While harvesting the compost, take care that you don't hurt the worms.

Quality of Organic Fertilizers

When you prepare compost or organic fertilizers, you can add a variety of organic matter to make the compost. The quality or nutrient value of the compost depends on what organic matter you add to it. Also, if you are purchasing organic fertilizers, you should try to know the nutrient content of the organic fertilizers before purchasing them. Just like chemical fertilizers, organic fertilizers also have an N-P-K (nitrogen potassium and phosphorus) value.

The NPK value for organic fertilizers depends on the organic matter used for preparing them. Organic matter such as dried blood and fish meal are rich in nitrogen. Fish meal is also rich in phosphorus. Kelp, on the other hand, is rich in potassium. The ability of different types of organic matter to release nutrients is also not the same. Therefore, the availability of nutrients is also an important criterion when selecting the type of organic fertilizer.

Compost

Compost is decomposed organic material. Compost is made with material such as leaves, shredded twigs, and kitchen scraps from plants. To gardeners, compost is considered "black gold" because of its many benefits in the garden. Compost is a great material for garden soil. Adding compost to clay soils makes them easier to work and plant. In sandy soils, the addition of compost improves the water holding capacity of the soil. By adding organic matter to the soil, compost can help improve plant growth and health.

Composting is also a good way to recycle leaves and other yard waste. Instead of paying a company to haul away leaves, you can compost the leaves and return the nutrients to your garden. Instead of buying peat moss, save money and make your own compost.

Composting Process

Composting is nature's way of recycling. Composting biodegrades organic waste. i.e. food waste, manure, leaves, grass trimmings, paper, wood, feathers, crop residue etc., and turns it into a valuable organic fertilizer.

Composting is a natural biological process, carried out under controlled aerobic conditions (requires oxygen). In this process, various microorganisms, including bacteria and fungi, break down organic matter into simpler substances. The effectiveness of the composting process is dependent upon the environmental conditions present within the composting system i.e. oxygen, temperature, moisture, material disturbance, organic matter and the size and activity of microbial populations.

Composting is not a mysterious or complicated process. Natural recycling (composting) occurs on a continuous basis in the natural environment. Organic matter is metabolized by microorganisms and consumed by invertebrates. The resulting nutrients are returned to the soil to support plant growth.

Composting is relatively simple to manage and can be carried out on a wide range of scales in almost any indoor or outdoor environment and in almost any geographic location. It has the potential to manage most of the organic material in the waste stream including restaurant waste, leaves and yard wastes, farm waste, animal manure, animal carcasses, paper products, sewage sludge, wood etc. and can be easily incorporated into any waste management plan.

Since approximately 45 - 55% of the waste stream is organic matter, composting can play a significant role in diverting waste from landfills thereby conserving landfill space and reducing the production of leachate and methane gas. In addition, an effective composting program can produce a high quality soil amendment with a variety of end uses.

The essential elements required by the composting microorganisms are carbon, nitrogen, oxygen and moisture. If any of these elements are lacking, or if they are not provided in the proper proportion, the microorganisms will not flourish and will not provide adequate heat. A composting process that operates at optimum performance will convert organic matter into stable compost that is odor and pathogen free, and a poor breeding substrate for flies and other insects. In addition, it will significantly reduce the volume and weight of organic waste as the composting process converts much of the biodegradable component to gaseous carbon dioxide.

The composting process is carried out by three classes of microbes:

- Psychrophiles - Low temperature microbes.

- Mesophiles - Medium temperature microbes.
- Thermophiles - High temperature microbes.

Generally, composting begins at mesophilic temperatures and progresses into the thermophilic range. In later stages other organisms including Actinomycetes, Centipedes, Millipedes, Fungi, Sowbugs, Spiders and Earthworms assist in the process.

Temperature

Temperature is directly proportional to the biological activity within the composting system. As the metabolic rate of the microbes accelerates, the temperature within the system increases. Conversely, as the metabolic rate of the microbes decreases, the system temperature decreases. Maintaining a temperature of 130 °F or more for 3 to 4 days favors the destruction of weed seeds, fly larvae and plant pathogens.

At a temperature of 155 degrees F, organic matter will decompose about twice as fast as at 130 degrees F. Temperatures above 155 degrees F may result in the destruction of certain microbe populations. In this case temperature may rapidly decline. Temperature will slowly rise again as the microbe population regenerates.

Moisture content, oxygen availability, and microbial activity all influence temperature. When the pile temperature is increasing, it is operating at optimum performance and should be left alone. As the temperature peaks, and begins to decrease, the pile should be turned to incorporate oxygen into the compost. Subsequently, the pile should respond to the turning and incorporation of oxygen, and temperature should again cycle upwards. The turning process should be continued until the pile fails to re-heat. This indicates that the compost material is biologically stable.

Composting microorganisms thrive in moist conditions. For optimum performance, moisture content within the composting environment should be maintained at 45 percent. Too much water can cause the compost pile to go anaerobic and emit obnoxious odors. Too little will prevent the microorganisms from propagating.

Particle Size

The ideal particle size is around 2 to 3 inches. In some cases, such as in the composting of grass clippings, the raw material may be too dense to permit adequate air flow or may be too moist. A common solution to this problem is to add a bulking agent (straw, dry leaves, paper, cardboard) to allow for proper air flow. Mixing materials of different sizes and textures also helps aerate the compost pile.

Turning

During the composting process oxygen is used up quickly by the microbes as they metabolize the organic matter. As the oxygen becomes depleted the composting process slows and temperatures decline. Aerating the compost by turning should ensure an adequate supply of oxygen to the microbes.

Composting Period

The composting period is governed by a number of factors including, temperature, moisture, oxygen, particle size, the carbon-to-nitrogen ratio and the degree of turning involved. Generally, effective management of these factors will accelerate the composting process.

Carbon to Nitrogen Ratio

The microbes in compost use carbon for energy and nitrogen for protein synthesis. The proportion of these two elements required by the microbes averages about 30 parts carbon to 1 part nitrogen. Accordingly, the ideal ratio of Carbon to Nitrogen (C:N) is 30 to 1 (measured on a dry weight basis). This ratio governs the speed at which the microbes decompose organic waste. Most organic materials do not have this ratio and, to accelerate the composting process, it may be necessary to balance the numbers.

Example, if you have two bags of cow manure (C:N = 20:1) and one bag of corn stalks (C:N = 60:1) then combined you have a C:N ration of (20:1 + 20:1 + 60:1)/3 = (100:1)/3 = 33:1.

Composters for Smaller Volumes

- Plastic bin (well ventilated).
- Metal or plastic drum (base removed – well ventilated).

Composters for Larger Volumes

Rotating drum (in vessel) Enclosure (made from 4 x 4 pallets lined with chicken wire) Open pile – windrow (covered with plastic or tarp).

In-vessel

An in-vessel, aerobic mechanical composter can be constructed from a steel drum, or tank designed to rotate at three to five revolutions per hour. Rotation can be carried out with a simple hand crank or a timed electrical mechanical device. This type of composter can produce a stabilized compost in three to four days and can be an environmentally appropriate, low management alternative to bin composting.

Aerated Bin

An aerated bin can be constructed using 4 × 4 pallets fastened together to form a box and lined with wire mesh. To limit the degree of turning and permit air to flow through the pile the structure can be elevated or, in the alternative, perforated pipes can be incorporated into the structure. One side of the structure should be detachable to facilitate loading, mixing and unloading. The composter should be waterproof and located in and area that is protected from the wind.

Static compost piles and windrows should be large enough to retain heat and small enough to facilitate air to its center. As a rule of thumb, the minimum dimensions of a pile should be 3 feet by 3 feet by 3 feet.

Turning Units

Turning units are ideally suited for batch composting and are extremely practical for building and turning active compost. Turning units allow convenient mixing for aeration and accelerated composting.

Composting Methods

Hot Composting

Hot composting is the most efficient method for producing quality compost in a relatively short time. In addition, it favors the destruction of weed seeds, fly larvae and pathogens. While hot composting, using the windrow or bin method, requires a high degree of management, hot composting, using the in-vessel method, requires a lesser degree of management.

Cold Composting

This method is ideal for adding organic matter around trees, in garden plots, in eroded areas etc. The time required to decompose organic matter using this method is governed, to a large extent, by environmental conditions and could take two years or more.

Sheet Composting

Sheet composting is carried out by spreading organic material on the surface of the soil or untilled ground and allowing it to decompose naturally. Over time, the material will decompose and filter into the soil. This method is ideally suited for forage land, no-till applications, erosion control, roadside landscaping etc. The process does not favor the destruction of weed seeds, fly larvae, pathogens etc. and composting materials should be limited to plant residue and manure. Again, decomposition time is governed by environmental conditions and can be quite lengthy.

Trench Composting

Trench composting is relatively simple. Simply dig a trench 6 - 8 inches deep, fill with 3 - 4 inches of organic material and cover with soil. Wait a few weeks and plant directly above the trench. This method does not favor the destruction of weed seeds, fly larvae and pathogens and the composting process can be relatively slow.

Loading the Bin/Windrow

Place the raw materials in layers using a balance of high carbon (moist) and low carbon (dry) materials. Each layer should be no more than four to six inches in depth. Spray each layer with a light mist of CBCT Stock Solution (Mix CBCT Concentrate and water at a rate of 1:200). This will initiate and accelerate the composting process and eliminate odors).

- Step 1. Start with a 4 to 6 inch layer of coarse material set on the bottom of the composter or on top of the soil.
- Step 2. Add a 3 to 4 inch layer of low carbon material.
- Step 3. Add a 4 to 6 inch layer of high carbon material.
- Step 4. Add a 1 inch layer of garden soil or finished compost.
- Step 5. Mix the layers of high carbon material, low carbon material, and soil or compost.

Repeat steps 2 through 5 until the composting bin is filled (maximum 4 feet in height). Cap with dry material.

Loading the Vessel (In-vessel Composting)

To accelerate the composting process, simply mix the high carbon and low carbon materials together before placing them in the composter. Add the mixture to the composter in small batches, spraying each batch with a light mist of water or CBCT stock solution.

Adding Material during the Composting Process

Ideally, new materials should be added to the composting system during turning or mixing. Generally, the addition of moist materials accelerates the composting process while the addition of dry materials slows the process.

About Compost

Finished compost can be classified as a 100% organic fertilizer containing primary nutrients as well as trace minerals, humus and humic acids, in a slow release form. Compost improves soil porosity, drainage and aeration and moisture holding capacity and reduces compaction. Compost can retain up to ten times it's weight in water. In addition, compost helps buffer soils against extreme chemical imbalances; aids in unlocking soil minerals; releases nutrients over a wide time window; acts as a buffer against the absorption of chemicals and heavy metals; promotes the development of healthy root zones; suppresses diseases associated with certain fungi; and helps plants tolerate drought conditions.

Things that can be Composted

Almost anything that comes from the ground can be composted. Cucumber ends, apple cores, carrot peels, cantaloupe rinds, avocado pits, an old pumpkin leftover from Halloween — any vegetable or fruit scrap will do. But grains also sprout from soil, which means you can throw stale bread, cereal, and pasta in your compost heap, too.

What else goes into the pile? Coffee grounds and filters, tea leaves, herbs, spices, and nut and egg shells are all acceptable ingredients. You can also include cut flowers or plant trimmings, as long as they aren't diseased.

But there's a long list of things you can't compost — and they mostly apply to animal products. No meat, fish, butter, yogurt, cheese, milk, or animal fat is allowed. It's also a good rule of thumb to keep anything oily or greasy out of the pile, and please, don't include your pet's poop. It may be organic, but it introduces disease to the compost that make it unusable.

Positive Impacts of Compost on the Environment

When the food in your trash goes to a landfill, it emits methane, a greenhouse gas that contributes to climate change. Composting is a way to minimize those methane emissions, but it does a lot more than that. Because compost is used as a fertilizer, it reduces the need for chemical options, which can be harsh on the earth. According to the EPA, compost is also capable of capturing and eliminating 99.6 percent of volatile organic compounds, or VOCs, from the air. These vapors and gases can have a harmful impact on your health, with side effects ranging from nausea to throat irritation.

Studies further show that compost can aid in carbon sequestration. When applied to soil, compost potentially functions as a "carbon sink," trapping and containing the element in the dirt. And if the carbon is in the ground, it isn't in our atmosphere, where it can wreak havoc on the planet.

Composting

Composting involves a lot of collecting — into a pile, heap, or bin. Instead of throwing out your coffee grounds or veggie scraps throughout the week, you just load them into your compost. The rest of the rules depend on where you're keeping that compost: indoors or outside.

Composting at Home

Let's say you're composting in a house with outdoor space. First, designate a spot outside for your compost heap. Some people maintain a literal heap of stuff on the ground, while others keep their items contained in a large tub. A tub is often easier for beginners, but if you're going with a heap, put up a small fence to keep critters out.

As you collect items in your compost, aim for a mix of nitrogen-rich "green" material and carbon-heavy "brown" material. Green materials tend to be wet, and include your fruit and vegetable scraps, coffee grounds, and tea leaves. Brown materials are drier, and span sawdust, newspaper clippings, and trimmings from dead plants.

Every other week, turn your compost pile with a shovel to let it aerate. Add water to keep it moist, and keep an eye out for any worrisome pests or disease. But otherwise, you can sit back and let nature do the rest of the work. It will take months for your heap of kitchen scraps to morph into soil, but you should notice a change after a few weeks.

Composting in an Apartment

Now let's say you live in a fourth-floor walk-up without any green space. You can still compost — and in fact, it may be even less labor-intensive than the backyard method.

Many cities now offer compost collection at farmers markets or designated drop sites. To participate, simply store your scraps in a container on your kitchen countertop or in your freezer. Then, on collection day, bring them to the market and dump them in communal compost tub.

Some cities also collect compost through trash and recycling services, often in a designated bin. Community gardens are another potential avenue, as they often accept or outright solicit compost from people in the neighborhood.

But if you have no city or community garden support, there's still vermicomposting. This self-contained system is designed for the indoors, but it often turns people off because it involves worms. You're essentially replicating what would happen naturally outside in a plastic bin with food scraps and redworms — as the Brooklyn Botanical Garden explains, you fill your bin with organic "bedding" like shredded newspaper or leaves, mix in a little soil, add the worms, and then feed them your compost. Keep a lid on the bin and to avoid any odors, store the whole thing in your fridge.

Can Compost go Bad?

Compost doesn't really "go bad," but it can get too dry, too wet, or too old. If it's too dry, add a little water to improve the moisture. If it's too wet, turn and mix all the ingredients, adding dry organic material like sawdust to absorb some of the extra liquid. And when compost is too old or otherwise "spent," it's not bad per se — it's just lost some nutrients. You can fix this by adding it into a fresh new pile of compost.

The Benefits of Composting

On top of all the great things compost does for the environment, it's also a valuable tool for farmers and gardeners. Compost boosts water retention in soil, which means the budding plants in that soil need less irrigation. It also tends to facilitate bigger crop yields, giving food producers a better harvest. This all adds up to more growth with less water — and, in turn, a more affordable way to create food, flowers, and other plants.

Mixed Fertilizers

Types of Mixed Fertilisers

A mixed fertiliser means a mixture of two or more straight fertilisers. For example, ammonium sulphate and single super phosphate may be thoroughly mixed to get a mixed fertiliser. Mixed fertilisers are marketed as wheat fertiliser mixtures, paddy fertiliser mixtures etc.

Mixed fertilisers are of two kinds:

- Open formula mixed fertiliser: In this case, the manufacturing firms disclose the names and quantities of the straight fertilisers that are constituents of the mixed fertilizer.
- Close formula mixture: In this case, the manufacturing firms do not disclose the names and quantities of the straight fertilisers that are constituents of the mixed fertiliser.

Materials used to Make Mixed Fertilisers

The following materials are required to make the mixed fertiliser:

- Straight fertilisers are the primary materials used for making the mixed fertiliser. They supply the fertiliser element required in the proper amount to the soil.
- Some low grade organic materials are added 100 pounds per ton to the mixed fertiliser in order to prevent caking, and to maintain them in a good physical condition. These organic materials (peat, paddy husk etc.) are called conditioners.
- Some basic material like Dolomite is added to neutralize acidity if the fertiliser mixture produces acids.
- Some materials like finely ground ash, soil etc. are added to make up the difference between the weight of the fertiliser necessary to supply the plant nutrients and the desired quantity of fertiliser mixture.

Incompatibilities in Fertiliser Mixture

A detailed knowledge of the properties of fertilisers is essential for preparing the fertiliser mixture.

The following kind of fertilisers should not be mixed at all or only mixed just before use:

- Ammonium sulphate, ammonium nitrate etc. should not mixed with basically reactive fertilisers like basic slag, rock phosphate etc., because they may decompose to liberate ammonia gas.

  $$(NH_4)_2SO_4 + CaCO_3 \rightarrow (NH_4)_2CO_3 + CaSO_4$$
  $$(NH_4)_2CO_3 \rightarrow 2NH_3 + CO_2 + H.OH$$

- Fertilisers like single super phosphate, ammonium phosphate etc. which contain water soluble phosphorus should not be mixed with fertilisers then contain free lime because water soluble phosphorus would be converted to water insoluble phosphorus as shown below:

  $$Ca(H_2PO_4)_2 + 2CaCO_3 \rightarrow Ca_3(PO_4)_2 + 2CO_2 + 2H_2O$$

- Some hygroscopic fertilisers like urea and sulphate of potash tend to form lumps after mixing. Therefore, they should be mixed only just before application.
- If urea is to be mixed with a phosphatic fertiliser, it should be mixed with mono or di-ammonium phosphate, or ammoniated single super phosphate or ammoniated triple super phosphate.
- Urea should not be mixed with ammonium nitrate because the mixture will readily liquefy.

Granulation of Mixed Fertiliser

Recently the granulated mixed fertilisers are being manufactured to decrease the separation of the constituent straight fertilisers and to prevent them from caking. There are more conveniently applied in the field and are difficult to adulterate. Most urea is either pilled or pelleted in sizes too large to form stable dry formulations with most other fertilisers. At present, a granular form of urea is also available.

Advantages of Mixed Fertilisers

- Two or more fertiliser elements are added together to make a mixed fertiliser to be applied to the field. Less labour is therefore required for the application of a mixed fertiliser.
- The fertiliser elements can be more uniformly applied to the field especially when they are required in small quantities.
- Mixed fertilisers can be easily drilled in the filed because they are in a good physical condition.
- When a mixed fertiliser has been prepared on a scientific basis and compensates for the deficiency of nutrients in the soils of the region, the farmer is assured of the right proportion of plant nutrients in the soil.

Disadvantages of a Mixed Fertiliser

- The use of mixed fertiliser does not permit application of individual fertiliser elements, which may be required by the crop at specific times.
- The illiterate farmer cannot effectively control the quantity of plant food present in the mixture. He has to rely entirely on the grade advertised by the firms.
- The firms charge for mixing the fertilisers.

Micronutrient Fertilizers

Beyond the basic three essential nutrients, Nitrogen, Phosphorous and Potassium (NPK), other requirements for plant growth and soil health are the secondary nutrients (S, Ca, Mg) and also micronutrients, which include iron, manganese, zinc, boron, molybdenum and copper. These secondary micronutrients are commercially available for agricultural application in many forms, including as enhancements to commodity forms of N, P and K fertilizers. Homogenous, granular micronutrient compounds are also produced and used either in direct application or as relatively minor components of bulk NPK blends. Secondary nutrients, especially sulfur, are often supplied in prilled or pastillated form. High magnesium and calcium analysis products are also commercially available in granular form.

Like any particulate fertilizer materials, micronutrient fertilizers can physically break down and generate dust as they age and undergo multiple handling steps during production, storage, transportation and application. An appropriately designed dust control coating and application system can minimize the problems associated with that dust generation and product deterioration.

Micronutrient Nutrition

Through increased use of soil testing and plant analyses, micronutrient deficiencies have been verified in many soils. Some reasons limiting the incidental additions of micronutrients include:

- High-yield crop demands remove micronutrients from the soil.
- Increased use of high-analysis NPK fertilizers containing lower quantities of micronutrient contaminants.
- Advances in fertilizer technology reduce the residual addition of micronutrients.

These factors contribute to the significant increase in usage of and need for micronutrients in order to achieve full balanced nutrition.

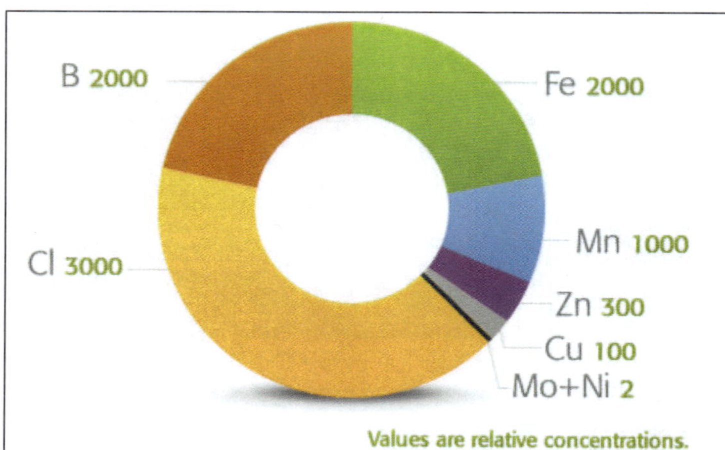

Micronutrients are just as important as macronutrients, but the amount required is very small.

Boron

Boron (B) exists primarily in soil solutions as the BO_3^{-3} anion– the form commonly taken up by plants. One of the most important micronutrients affecting membrane stability, B supports the structural and functional integrity of plant cell membranes. Boron-deficiency symptoms first appear at the growing points, and certain soil types are more prone to boron deficiencies.

Copper

Copper (Cu) activates enzymes and catalyzes reactions in several plant-growth processes. The presence of copper is closely linked to Vitamin A production, and it helps ensure successful protein synthesis.

Iron

Iron (Fe) is essential for crop growth and food production. Plants take up Fe as the ferrous (Fe^{2+}) cation. Iron is a component of many enzymes associated with energy transfer, nitrogen reduction and fixation, and lignin formation.

Manganese

Manganese (Mn) functions primarily as part of enzyme systems in plants. It activates several important metabolic reactions and plays a direct role in photosynthesis. Manganese accelerates germination and maturity while increasing the availability of phosphorus (P) and calcium (Ca).

Molybdenum

Molybdenum (Mo) is a trace element found in the soil, and is required for the synthesis and activity of the enzyme nitrate reductase. Molybdenum is vital for the process of symbiotic nitrogen (N) fixation by Rhizobia bacteria in legume root modules. Considering molybdenum's importance in optimizing plant growth, it's fortunate that Mo deficiencies are relatively rare in most agricultural cropping areas.

Zinc

Zinc (Zn) is taken up by plants as the divalent Zn^{+2} cation. It was one of the first micronutrients recognized as essential for plants and the one most commonly limiting yields. Although Zn is required only in small amounts, high yields are impossible without it.

Chlorine

Plants take up chlorine Cl as the chloride Cl^- anion. It's active in energy reactions in the plant. Most Cl^- in soils comes from salt trapped in parent materials, marine aerosols and volcanic emissions. Classified as a micronutrient, Cl^- is required by all plants in small quantities.

Nickel

Nickel (Ni) was added to the list of essential plant nutrients late in the 20th century. Nickel is important in plant N metabolism because it is a component of the urease enzyme. Without the presence of Ni, urea conversion is impossible. It is required in very small amounts, with the critical level appearing to be about 1.1 ppm.

Application with Mixed Fertilizers

The most common method of micronutrient application for crops is soil application. Recommended application rates usually are less than 10 lb/acre (on an elemental basis), so uniform application of micronutrient sources separately in the field is difficult. Therefore, both granular and fluid NPK fertilizers are commonly used as carriers of micronutrients. Including micronutrients with mixed fertilizers is a convenient method of application, and allows more uniform distribution with conventional application equipment. Costs are also reduced by eliminating a separate application. Four methods of applying micronutrients with mixed fertilizers are:

- Incorporation with granular fertilizers: Incorporation during manufacture results in uniform distribution of micronutrients throughout granular NPK fertilizers.
- Bulk blending with granular fertilizers: Bulk blending produces fertilizer grades that provide the recommended micronutrient rates. Unfortunately, nutrient segregation is common, resulting in uneven nutrient distribution.
- Coating onto granular fertilizers: Coating powdered micronutrients onto granular NPK fertilizers decreases the possibility of segregation.

- Mixing with fluid fertilizers: Mixing micronutrients with fluid fertilizers has become a popular method of application. Keep in mind, compatibility tests should be made before tank-mixing operations of micronutrients with fluid fertilizers are attempted. Suspension fertilizers are used as micronutrient carriers as well.

Foliar Sprays

Foliar sprays are widely used to apply micronutrients, especially iron and manganese, for many crops. Soluble inorganic salts generally are as effective as synthetic chelates in foliar sprays, so the inorganic salts usually are chosen because of lower costs. Suspected micronutrient deficiencies may be diagnosed with foliar spray trials with one or more micronutrients, but tissue sampling is the most common method to determine deficiencies during the growing season. Correction of deficiency symptoms usually occurs within the first several days, and then the entire field could be sprayed with the appropriate micronutrient source. Inclusion of sticker-spreader agents in the spray is suggested to improve adherence of the micronutrient source to the foliage. Caution should be used because of leaf burn due to high salt concentrations or inclusion of certain compounds in foliar sprays.

Advantages of Foliar Sprays

- Application rates are much lower than rates for soil application.
- Uniform application is easily obtained.
- Response to the applied nutrient is almost immediate, so deficiencies can be corrected during the growing season.

Disadvantages of Foliar Sprays

- Leaf burn may result if salt concentrations of the spray are too high.
- Nutrient demand often is high when the plants are small and leaf surface is insufficient for foliar absorption.
- Maximum yields may not be possible if spraying is delayed until deficiency symptoms appear.
- There is little residual effect from foliar sprays.
- Application costs are higher if more than one spray is needed unless they can be combined with pesticide spray applications.

Micronutrient Rates

Boron

Recommended application rates of boron are rather low (0.5 to 2 lb/acre), but should be carefully followed because the range between boron deficiency and toxicity in most plants is narrow. Uniform application of boron in the field is very important for the above reason. Boronated NPK fertilizers (those containing boron sources incorporated at the factory) will ensure a more uniform application than most bulk blended fertilizers. Foliar sprays also ensure a rather uniform application, but costs generally are higher.

Soil tests should be included in boron fertilization programs, first to assess the level of available boron, and later to determine possible residual effects (buildup). The most common soil test for boron is the hot-water-soluble test. This test is more difficult to conduct than most other micronutrient soil tests, but most boron response data have been correlated with it.

Copper

Recommended copper rates range from 3 to 10 lb/acre as $CuSO_4$ or finely ground CuO. Residual effects of applied copper are very marked, with responses being noted up to eight years after application. Because of these residual effects, soil tests are essential to monitor possible copper accumulations to toxic levels in soils where copper fertilizers are being applied.

Plant analyses also can be used to monitor copper levels in plant tissues. Copper applications should be decreased or discontinued when available levels increase beyond the deficiency range.

Iron

Soil applications of most iron sources generally are not effective for crops, so foliar sprays are the recommended application method. Spray applications of a 3 to 4% $FeSO_4$ solution at 20 to 40 gallons/acre are used to correct iron deficiencies. The application rate should be high enough to wet the foliage.

More than one foliar application may be required for correction of iron chlorosis. Inclusion of a sticker-spreader agent in the spray is suggested to improve adherence of the spray to the plant foliage for increased iron absorption by the plant.

Manganese

Recommended application rates range from 2 to 20 lb/acre of manganese, generally as $MnSO_4$. Application rates of MnO would be similar if applied as a fine powder or in NPK fertilizers. Band application of manganese sources with acid-forming fertilizers results in a more efficient use of applied manganese because the rate of oxidation of applied manganese to the unavailable tetravalent form (as in MnO_2) is decreased.

There are no residual effects of applied manganese for the same reason, so annual applications are needed. Foliar spray applications of $MnSO_4$ also are used and require lower rates than soil applications.

Molybdenum

Recommended molybdenum rates are much lower than those for the other micronutrients, so uniform application is very important. Broadcast application of molybdenized phosphate fertilizers prior to planting or to pastures has been used to correct molybdenum deficiencies. Soluble molybdenum sources also can be sprayed on the soil surface before tillage to obtain a uniform application.

Seed treatment is the most common method of molybdenum application. Molybdenum sources are coated onto the seed with a sticking agent and/or conditioner. This method ensures a uniform application, and sufficient amounts of molybdenum can be seed coated to provide sufficient molybdenum.

Zinc

Recommended rates of zinc generally range from 1 to 10 lb/acre. Band or broadcast applications are used, but foliar applications also are effective. Band applications of zinc sources with starter fertilizers is a common practice for row crops. Foliar sprays of a 0.5% $ZnSO_4$ solution applied at a rate of 20 to 30 gallons/acre also will supply sufficient zinc, but several applications may be necessary.

As with copper, residual effects of applied zinc are substantial, with responses found at least 5 years after application. Because of these residual effects, soil test levels of available zinc generally increase after several applications. Many states have reduced their recommended zinc application rates because of these residual effects.

Secondary Nutrient Fertilizers

Calcium, magnesium, and sulfur are essential plant nutrients. They are called "secondary" nutrients because plants require them in smaller quantities than nitrogen, phosphorus, and potassium. On the other hand, plants require these nutrients in larger quantities than the "micronutrients" such as boron and molybdenum.

Calcium, magnesium, and sulfur are generally adequate in most Mississippi soils with favorable pH and organic matter levels. They affect pH when applied to the soil. Calcium and magnesium both increase soil pH, but sulfur from some sources reduces soil pH. Compounds containing one or more of these nutrients are often used as soil amendments rather than strictly as suppliers of plant nutrition.

Calcium

Calcium is a low-key essential nutrient that carries a heavy load in plant growth. Too often, it takes a backseat as soil fertility programs are developed for many high-yield and high-quality crops. Peanut and tomato growers are probably exceptions in their emphasis on good calcium nutrition.

Functions of Calcium in Soil

In soil, calcium replaces hydrogen (H) ions from the surface of soil particles when limestone is added to reduce soil acidity. It is essential for microorganisms as they turn crop residues into organic matter, release nutrients, and improve soil aggregation and water holding capacity. Calcium helps enable nitrogen-fixing bacteria that form nodules on the roots of leguminous plants to capture atmospheric nitrogen gas and convert it into a form that plants can use.

Functions of Calcium in Plants

Calcium, along with magnesium and potassium, helps to neutralize organic acids, which form during cell metabolism in plants. Calcium also plays a role in other key plant functions:

- Improves the absorption of other nutrients by roots and their translocation within the plant.
- Activates a number of plant growth-regulating enzyme systems.
- Helps convert nitrate-nitrogen into forms needed for protein formation.
- Is needed for cell wall formation and normal cell division.
- Improves disease resistance.

Calcium Deficiency

Calcium deficiencies are most likely to occur in acid, sandy soils from which calcium has been leached by rain or irrigation water. It may also occur in strongly acid peat and muck soils where total calcium is low.

Calcium deficiency is not likely for most crops when the soil is properly limed to adjust soil pH to optimum levels for crop production. As soils become more acidic, crop growth is often restricted by toxic soil concentrations of aluminum and/or manganese — not a calcium shortage. Soil testing and a good liming program are the best management practices (BMPs) to prevent these problems.

Calcium deficiency can be prevented by following several BMPs such as soil testing on a regular basis and correcting soil acidity with proper liming. Balance the plant nutrition program by keeping calcium, potassium and magnesium available in a balanced supply. An overabundance of one can lead to a shortage or uptake (antagonism) of another. Also apply calcium for specific plant functions. For example, calcium applied when peanuts begin to set pods can help improve seed.

Symptoms of Calcium Deficiency

New leaf growth may slow and leaf tips may stick together. Remember that calcium does not readily translocate within the plant so deficiency symptoms will appear on the new growth.

Sources of Calcium

A good liming program is an efficient supplier of calcium to most crops. High-quality calcitic limestone is effective when pH adjustments are needed. If magnesium is deficient also, dolomitic limestone may be used, or calcitic limestone may be applied along with a magnesium source such as potassium-magnesium-sulfate. Gypsum (calcium sulfate) provides calcium when soil pH is adequate. Some common sources of calcium are shown at right.

Common Calcium Sources		
Material	% CA	Acid-Neutralizing
Gypsum	22%	—
Basic Slag	29%	50–70
Calcitic Limestone	32%	85–100
Dolomitic Limestone	22%	95–108
Hydrated Lime	46%	120–135
Pure calcium carbonate=100		

Magnesium

Energy is required for proper plant growth. Wheat and other crops require magnesium to capture the sun's energy for growth and production through photosynthesis. Magnesium is an essential component of the chlorophyll molecule, with each molecule containing 6.7 percent magnesium. Chlorophyll, the green pigment in plants, is the site where photosynthesis occurs. Without chlorophyll, plants could not manufacture food, and life on Earth would cease to exist.

Magnesium also acts as a phosphorus carrier in plants. It is necessary for cell division and protein formation. Phosphorus uptake could not occur without magnesium, and vice versa. So, magnesium is essential for phosphate metabolism, plant respiration and the activation of several enzyme systems.

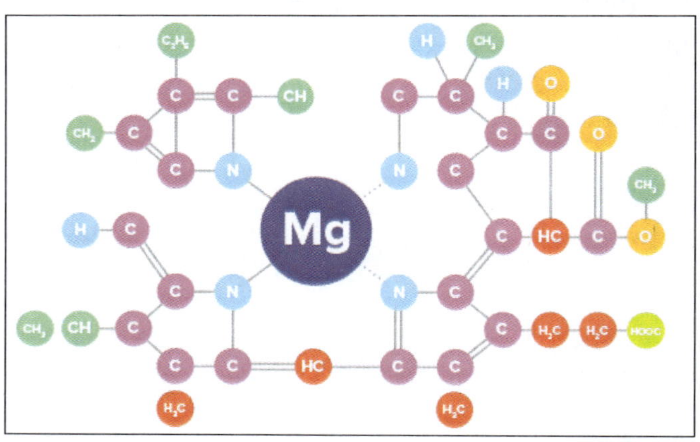

Hidden in the heart of each molecule of chlorophyll is an atom of magnesium.

Magnesium in Soils

The Earth's crust contains about 1.9 percent Mg, largely in the form of Mg-containing minerals. As these minerals slowly weather, some Mg is made available to plants. The supply of available Mg has been or is being depleted in some soils through leaching, plant uptake and removal processes.

Magnesium availability to plants is often related to soil pH. Research has shown that Mg availability to the plant decreases at low pH values. On acid soils with a pH below about 5.8, excessive hydrogen and aluminum can influence Mg availability and plant uptake. At high pH values (above 7.4), excessive calcium may have an overriding influence on Mg uptake by plants.

Sandy soils with low cation exchange capacity have a low Mg-supplying power.

Application of high-calcium limestone can aggravate an Mg deficiency by increasing plant growth and increasing the demand for Mg. High applications of ammonium and potassium may also interfere with balanced nutrition through competitive ion effects. If soil test levels are below 25 to 50 parts per million (ppm) – 50 to 100 lb/acre – exchangeable Mg is usually considered low, and Mg application is warranted.

A Rule of Thumb

For soils with a cation exchange capacity (CEC) higher than about 5 milliequivalents (ME) per 100 grams, it may be desirable to maintain the soil Ca-to-Mg ratio at about 10 to 1. For sandy soils with a CEC of 5 ME or less, it may be desirable to maintain the Ca-to-Mg ratio at about 5 to 1.

Symptoms of Magnesium Deficiency

Magnesium is mobile within the plant and easily translocated from older to younger tissues. When deficiencies occur, the older leaves are affected first.

How to Treat Mg Deficiencies

If Mg deficiencies are detected in growing crops through plant tissue analyses, a soluble magnesium source may be applied and watered into the soil by irrigation or rainfall. This will permit root access and plant uptake. Small amounts of Mg can also be applied to growing crops through foliar fertilization to correct or prevent developing deficiencies. The preferred approach is to soil-apply the required amounts of Mg before crops are planted or before they begin active growth.

Sources of Magnesium		
Material	% MG	Water Solubility
Dolomitic Limestone	6–12%	No
K-Mag	10–11%	Yes
Magnesium Chloride	7.5%	Yes
Magnesium Hydroxide	40%	No
Magnesium Nitrate	16%	Yes
Magnesium Oxide	56–60%	No
Magnesium Sulfate	10–16%	Yes

Sulfur

Sulfur in Soil

Sulfur is supplied to plants from the soil by organic matter and minerals, but it is often present in insufficient quantities and at inopportune times for the needs of many high- yielding crops. Most S in the soil is tied up in the organic matter and cannot be used by plants until it is converted to the sulfate (SO_4^{-2}) form by soil bacteria. That process is known as mineralization.

Sulfate is mobile in the soil, just as nitrate-nitrogen is mobile, and in some soils, it can be leached beyond the active root zone with heavy rainfall or irrigation. Sulfate may move back upward toward the soil surface as water evaporates, except in the sandier, coarse- textured soils that may be void of capillary pores. This mobility of sulfate-sulfur makes it difficult to calibrate soil tests and to use them as predictive tools for sulfur fertilization needs.

Sulfur tends to be held by clay soil particles more than nitrate-nitrogen. When early spring rains occur, soils with a sandy topsoil, but containing relatively high amounts of clay in the subsoil, may have sulfate-sulfur leached out of the topsoil but retained in the subsoil. Therefore, crops grown on these types of soils may show early S deficiency, but as the roots penetrate into the subsoil, the deficiency may disappear. On deep sandy soils with little or no clay in the subsoil, plants will likely respond to sulfur applications.

Sulfur in Plants

Sulfur is part of every living cell and required for synthesis of certain amino acids (cysteine and methionine) and proteins. Sulfur is also important in photosynthesis and crop winter hardiness. In addition, sulfur is important in the nitrate-reductase process, during which nitrate-nitrogen is converted to amino acids.

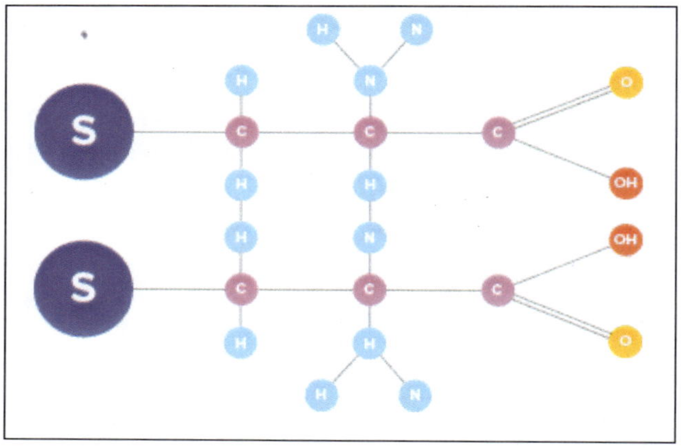

Sulfur is required for the synthesis of vitamins, and is a constituent of certain amino acids, which are the building blocks from which proteins are created. Without proteins, plants wither and die.

Sulfur Deficiency

In the field, sulfur deficiency and nitrogen deficiency are often easily confused. Symptoms of both deficiencies may appear as stunted plants, with a general yellowing of leaves. Sulfur is immobile within the plant and does not readily move from old to new growth. With sulfur deficiency, yellowing symptoms often first appear in younger leaves, whereas with nitrogen deficiency, the yellowing appears on the older leaves first. In less severe situations, visual symptoms may not be noticeable.

The best way to diagnose a deficiency is with a plant tissue analysis that includes an assay for both sulfur and nitrogen. Sulfur concentrations in most plants should range from about 0.2 to 0.5 percent. Desirable total nitrogen to total sulfur ratios have been considered, and range from 7:1 to 15:1. Wider ratios may point to possible sulfur deficiency, but should be considered along with actual N and S concentrations in making diagnostic interpretations.

When sulfur is deficient, nitrate-nitrogen may accumulate. This can pose significant health threats to grazing ruminants or those consuming hay high in nitrates. When nitrates accumulate in the plant, seed formation can be inhibited in some crops such as canola. Balancing sulfur with nitrogen nutrition is important to both plant and animal health.

Crops such as hybrid bermudagrass, alfalfa and corn that have a high dry- matter production generally require the greatest amount of sulfur. Also, potatoes and many other vegetables require large amounts of S, and have produced best when S is included in the fertility program. Without adequate S fertilization, crops that receive high rates of nitrogen may develop sulfur deficiencies.

Sulfur Sources

Some irrigation waters may contain significant quantities of sulfur. For example, when the irrigation water exceeds about 5 parts per million (ppm) sulfate S, a sulfur deficiency is unlikely. Most fertilizer sources of sulfur are sulfates and are moderately to highly soluble in water. The most important water-insoluble sulfur source is elemental S, which must be oxidized through bacterial action to the sulfate form before it can be utilized by plants. This oxidation is favored by warm soil temperatures, adequate soil moisture, soil aeration and fine sulfur particle sizes. If elemental sulfur is used, it should be incorporated into the soil well in advance of the crop needs.

References

- Soil-fertility-its-meaning-causes-and-maintenance-with-diagram, soil: biologydiscussion.com, Retrieved 23 April, 2019
- Uses-of-chemical-fertilizers: hunker.com, Retrieved 25 May, 2019

- Organic-fertilizer, soil-fertilizers, garden-how-to, gardeningknowhow.com, Retrieved 27 June, 2019
- Garden-how-to/soil-fertilizers/soil-fertilizers, garden: gardeningknowhow.com, Retrieved 29 July, 2019
- Benefits-of-organic-fertilizers, organic-cultivation, organic-products: organicfacts.net, Retrieved 2 August, 2019
- What-is-composting, food: greenmatters.com, Retrieved 4 January, 2019
- Mixed-fertilisers-types-granulation-and-advantages, fertilizers: soilmanagementindia.com, Retrieved 6 February, 2019
- Micronutrient-fertilizer, fertilizer-production, industries: arrmaz.com, Retrieved 8 March, 2019
- Micronutrients: cropnutrition.com, Retrieved 10 April, 2019
- Secondary-plant-nutrients-calcium-magnesium-and-sulfur: msstate.edu, Retrieved 12 May, 2019
- Secondary-nutrients: cropnutrition.com, Retrieved 14 June, 2019

Chapter 4

Important Fertilizers and their Manufacturing

Fertilizers play a major role in the enhancement and growth of plants. Nitrogen fertilizers, phosphorus fertilizers and potassium fertilizers are a few types of important fertilizers. The fertilizers made up of ammonia are known as nitrogen fertilizers. Phosphate fertilizers are obtained by extraction from minerals containing the phosphate ion. The topics elaborated in this chapter will help in gaining a better perspective about the different fertilizers and their manufacturing process.

Fertilizer

Fertilizer is a natural or artificial substance containing the chemical elements that improve growth and productiveness of plants. Fertilizers enhance the natural fertility of the soil or replace the chemical elements taken from the soil by previous crops.

A brief treatment of fertilizer follows. For full treatment, see agricultural technology: Fertilizing and conditioning the soil.

The use of manure and composts as fertilizers is probably almost as old as agriculture. Modern chemical fertilizers include one or more of the three elements that are most important in plant nutrition: nitrogen, phosphorus, and potassium. Of secondary importance are the elements sulfur, magnesium, and calcium.

Most nitrogen fertilizers are obtained from synthetic ammonia; this chemical compound (NH_3) is used either as a gas or in a water solution, or it is converted into salts such as ammonium sulfate, ammonium nitrate, and ammonium phosphate, but packinghouse wastes, treated garbage, sewage, and manure are also common sources of it. Phosphorus fertilizers include calcium phosphate derived from phosphate rock or bones. The more soluble superphosphate and triple superphosphate preparations are obtained by the treatment of calcium phosphate with sulfuric and phosphoric acid, respectively. Potassium fertilizers, namely potassium chloride and potassium sulfate, are mined from potash deposits. Mixed fertilizers contain more than one of the three major nutrients—nitrogen, phosphorus, and potassium. Mixed fertilizers can be formulated in hundreds of ways.

On modern farms a variety of machines are used to apply synthetic fertilizer in solid, gaseous, or liquid form. One type distributes anhydrous ammonia, a liquid under pressure, which becomes a nitrogenous gas when freed from pressure as it enters the soil. A metering device operates valves to release the liquid from the tank. Solid-fertilizer distributors have a wide hopper, with holes in the bottom; distribution is effected by various means, such as rollers, agitators, or endless chains traversing the hopper bottom. Broadcast distributors have a tub-shaped hopper from which the material falls onto revolving disks that distribute it in a broad swath.

Fertilizing and Conditioning the Soil

Soil fertility is the quality of a soil that enables it to provide compounds in adequate amounts and proper balance to promote growth of plants when other factors (such as light, moisture, temperature, and soil structure) are favourable. Where fertility of a soil is not good, natural or manufactured materials may be added to supply the needed plant nutrients; these are called fertilizers, although the term is generally applied to largely inorganic materials other than lime or gypsum. Fertilizer grade is a conventional expression that indicates the percentage of plant nutrients in a fertilizer; thus, a 10–20–10 grade contains 10 percent nitrogen, 20 percent phosphoric oxide, and 10 percent potash. The green plant, however, requires more nutrients than these.

Essential Plant Nutrients

In total, the plant has need of at least 16 elements, of which the most important are carbon, hydrogen, oxygen, nitrogen, phosphorus, sulfur, potassium, calcium, and magnesium.

The plant obtains carbon and hydrogen dioxide from the atmosphere; other nutrients are taken up from the soil. Although the plant contains sodium, iodine, and cobalt, these are apparently not essential. This is also true of silicon and aluminum.

Overall chemical analyses indicate that the total supply of nutrients in soils is usually high in comparison with the requirements of crop plants. Much of this potential supply, however, is bound tightly in forms that are not released to crops fast enough to give satisfactory growth. Because of this, the farmer is interested in measuring the available nutrient supply as contrasted to the total quantities.

The solid content of soils is broadly classified as organic and inorganic. Materials of organic origin range from fresh plant tissue to the more or less stable black or brown degradation product (humus) formed by biological decay. The organic matter is a potential source of nitrogen, phosphorus, and sulfur; it contains more than 95 percent of the total nitrogen, 5 to 60 percent of the total phosphorus, and 10 to 80 percent of the total sulfur. These three elements are cycled through the entire environment of living things (the biosphere). The soil organic matter can be considered as one of the storage points in these cycles. Where nonlegumes are grown in the absence of fertilizer or manures, the crop must gain its nitrogen supply from the organic matter; only a part, however, of the needed phosphorus and sulfur is so supplied.

The inorganic or mineral fraction, which comprises the bulk of most soils, is derived from rocks and their degradation products. The power to supply plant nutrients is much greater in the larger particles, sand and silt, than in the fine particles, or clay. The minerals that comprise the sand and silt in soil contain most of the elements essential for plant growth as a part of their structure. The difficulty is that these minerals decompose so slowly in soil that the rate of supply of the nutrient elements is usually insufficient for good growth of plants.

When the available supply of a given nutrient becomes depleted, its absence becomes a limiting factor in plant growth, and the addition of this nutrient to the soil will increase yields of dry matter. Excessive quantities of some nutrients may cause decrease in yield, however.

Determining Nutrient Needs

Determination of a crop's nutrient needs is an essential aspect of fertilizer technology. The appearance of a growing crop may indicate need of fertilizer; in some plants, however, the need for more or different nutrients may not be easily observable. If such a problem exists, its nature must be diagnosed, the degree of deficiency must be determined, and the amount and kind of fertilizer needed for a given yield must be found. There is no substitute for detailed examination of plants and soil conditions in the field, followed by simple fertilizer tests, quick tests of plant tissues, and analysis of soils and plants.

Sometimes plants show symptoms of poor nutrition. Chlorosis (general yellow or pale-green colour), for example, indicates lack of sulfur and nitrogen. Iron deficiency produces white or pale-yellow tissue. Symptoms can be misinterpreted, however. Plant disease can produce appearances resembling mineral deficiency, as can various organisms. Drought or improper cultivation or fertilizer application each may create deficiency symptoms.

After field diagnosis, the conclusions may be confirmed by experiments in a greenhouse or by making strip tests in the field. In strip tests, the fertilizer elements suspected of being deficient are added, singly or in combination, and the resulting plant growth observed. Next, it is necessary to determine the extent of the deficiency.

An experiment in the field can be conducted by adding nutrients to the crop at various rates. The resulting response of yield in relation to amount of nutrient supplied will indicate the supplying power of the unfertilized soil in terms of bushels or tons of produce. If the increase in yield is large, this practice will show that the soil has too little of a given nutrient. Such field experiments may not be practical, because they can cost too much in time and money. Soil-testing laboratories are available in most areas; they conduct chemical soil tests to estimate the availability of nutrients. Commercial soil-testing kits give results that may be very inaccurate, depending on techniques and interpretation. Actually, the most accurate system consists of laboratory analysis of the nutrient content of plant parts, such as the leaf. The results, when correlated with yield response to fertilizer application in field experiments, can give the best estimate of deficiency. Further development of remote sensing techniques, such as infrared photography, are under study and may ultimately become the most valuable technique for such estimates.

The Economics of Fertilizers

The practical goal is to determine how much nutrient material to add. Since the farmer wants to know how much profit to expect if he buys extra fertilizer, the tests are interpreted as an estimation of increased crop production that will result from nutrient additions. Cost of nutrients must be balanced against value of crop or even against alternative procedures, such as investing the money in something else with a greater potential return. The law of diminishing returns is well exemplified in fertilizer technology. Past a certain point, equal inputs of chemicals produce less and less yield increase. The goal of the farmer is to use fertilizer in such a way that the most profitable rate is employed.

Fertilizers can aid in making profitable changes in farming. Operators can reduce costs per unit of production and increase the margin of return over total cost by increasing rates of application of fertilizer on principal cash and feed crops. They are then in a position to invest in soil conservation and other improvements that are needed when shifting acreage from surplus crops to other uses.

With the global population steadily growing, it is important that enough crops are produced each year to provide food, clothing and other agricultural products to people around the world. Crops such as corn, wheat and cotton receive nutrients from the soil they are grown in; various crops deplete soil nutrients in different ways and rates. Some crop growth can deplete soil nutrients after just a few seasons of planting. Fertilizers play an important role in providing crops with the nutrients they need to grow and be harvested for nutritious food.

Fertilizers help deliver enough food to feed the world's population. But they can do even more. A class of fertilizers called micronutrient fertilizers is engineered to enrich crops with vital nutrients that help support human health. For example, micronutrients such as zinc are important to human nutrition, especially children. According to a United Nations study, much of the world's grain crops are grown in soil without adequate zinc; offering micronutrient fertilizers to grain crops enriches the grain with an important nutrient.

Safety Information

Fertilizers are essential to the security of the world's food supply, and they must be used properly. The manufacture, sale and transportation of fertilizers is heavily regulated. States have difference regulations and statutes that address fertilizer use and production to protect human health and the environment.

In addition to individual state regulations, the federal government also regulates fertilizer production, use and transport. Fertilizer producers are required to participate in the U.S. Occupational Safety and Health Administration's (OSHA) Process Safety Management Regulation to protect worker safety. In addition to OSHA, the U.S. Department of Agriculture, Environmental Protection Agency, Department of Homeland Security and Department of Transportation all play roles in regulating fertilizers and their production.

Ammonium nitrate, a good source of nitrogen and ammonium for plants, is an important ingredient in the production of high quality, effective fertilizers. Ammonium nitrate is produced using anhydrous ammonia, a gas that can become explosive when mixed with air. While this fertilizer is applied safely every day, this feature also means that it can be dangerous in the wrong hands. Fertilizer producers take great care in the safe use and storage of anhydrous ammonia and have implemented safety measures to help prevent or mitigate an incident.

Types of Fertilizers

There are different types of fertilizers that are mentioned below:

Inorganic Fertilizers

Inorganic fertilizers are chemical fertilizers that contain nutrient elements for the growth of crops made by chemical means.

The inorganic fertilizers are of the following types:

Nitrogen Fertilizers

Nitrogen fertilizers contain nitrogen necessary for the development of crops. Nitrogen is the main constituent of chlorophyll that maintains a balance in the process of photosynthesis. It is also a part of amino acids in plants and constitutes protein. Nitrogen fertilizers improve the production and quality of agricultural products.

Phosphorus Fertilizer

The main nutrient in a phosphorus fertilizer is phosphorus. The efficiency of fertilizer depends upon effective phosphorus content, methods of fertilizing, properties of soil and crop strains. Phosphorus found in the protoplasm of the cell plays an important role in cell growth and proliferation. The phosphorus fertilizer is beneficial for the growth of roots of the plants.

Organic Fertilizers

Organic fertilizers are natural fertilizers obtained from plants and animals. It enriches the soil with carbonic compounds essential for plant growth. Organic fertilizers increase the organic matter content of the soil, promotes the reproduction of microorganisms, and changes the physical and chemical properties of the soil. It is considered to be one of the main nutrients for green food.

Organic fertilizers can be obtained from the following products:

- Agricultural Waste.
- Livestock Manure.
- Industrial Waste.
- Municipal Sludge.

Advantages of Fertilizers

The advantages of fertilizers are mentioned below:

- They are easy to transport, store, and apply.
- For supplying a specific nutrient we can select a specific fertilizer due to its nutrient specific nature.
- They are water soluble and can easily dissolve in the soil. Hence, they are easily absorbed by the plants.
- They have a rapid effect on the crops.
- They increase the crop yield and provide enough food to feed the large population.
- They are predictable and reliable.

Disadvantages of Fertilizers

Fertilizers have the following disadvantages:

- They are expensive.
- The ingredients in the fertilizers are toxic to the skin and respiratory system.
- Excessive use of fertilizers damages the plants and reduces soil fertility.
- Leaching occurs and the fertilizers reach the rivers causing eutrophication.
- Long term use reduces the microbial activity and disturbs the pH of the soil.

Uses of Fertilizers

Fertilizers are used for various purposes. The uses of fertilizers are mentioned below:

- They are used to provide additional nutrients to the plants.
- They are added to improve the yield of the crops.
- Nitrogen-rich fertilizers are used for the greening of lawns.
- Organic fertilizers improve the texture and fertility of the soil.
- Gardeners use fertilizers to address certain needs of the plants such as nutritional needs.
- Fertilizers are added to potted plants to replace the lost nutrients.

Importance of Fertilizers

It is very difficult to meet the demands of the increasing population with such less resources. Loss of soil fertility, pests, and lack of nutrients have resulted in a decrease in agricultural production. This has increased the importance of fertilizers in agriculture.

Fertilizers can be important to the plants in the following ways:

- Fertilizers increase plants' tolerance towards pests. This has reduced their reliance on insecticides and herbicides, thereby, producing healthier crops. Consequently, diseases have reduced providing aesthetic value to the crops.
- Fertilizers improve the water holding capacity of the plants and increase root depth.
- The potassium content present in the fertilizers strengthens the straws and stalks of the plants.
- The phosphorus present in the fertilizers helps in the faster development of roots and formation of seeds in the plants.
- Nitrogen in the fertilizers enhances the growth of the plants which can be characterized by the green colour of the plants.

Since the chemical fertilizers adversely affect soil fertility, biofertilizers were brought into use. These are substances that contain living or latent cells, and even micro-organisms. They provide the soil with the necessary nutrients and microbes for the growth of the plants. They help the soil to retain its fertility. They are environment-friendly. They also destroy pathogenic components responsible for causing disease in plants. Acetobacter and Rhizobium are two such widely used biofertilizers.

Environmental Issues

A 'green' controversy exists between those who believe that no 'artificial' fertilizers should be used and those who believe that manufactured fertilizers have to be used in order to feed an ever growing world population. It is estimated that synthetic nitrogen fertilizers now provide about half the protein in crops worldwide.

Because manufactured fertilizers contain precisely the same plant nutrients in the chemical forms that are provided in nature, they are not in themselves environmentally harmful. However, it is important that the timing and rate of fertilizer or manure application are correct, calculated to optimize crop production and minimize adverse environmental pollution from nutrient losses.

The nitrate ion is particularly mobile in soil and problems occur if excessive amounts of nitrogen fertilizers or animal manures are used. This is because any unused nitrate remains in the soil and there is a risk that it will leach into the water courses.

Phosphate ions applied in fertilizer and manure become tightly bound to soil particles and are not normally leached. Some phosphate ions however are found in rivers and they come from soil which has been eroded into the water course.

Nevertheless, even very small amounts of phosphate transferred to water can lead to nutrient enrichment of the water - eutrophication - which stimulates plant growth. Decomposition of excessive plant growth requires large amounts of oxygen and this decreases the amount available for fish and other aquatic animals leading to their death and, in extreme cases, eventual conversion of a waterway into an anaerobic waterbody which is toxic to vertebrates and smells badly.

Nitrogen Fertilizers

Nitrogen in Plants

Healthy plants often contain 3 to 4 percent nitrogen in their above-ground tissues. This is a much higher concentration compared to other nutrients. Carbon, hydrogen and oxygen, nutrients that don't play a significant role in most soil fertility management programs, are the only other nutrients present in higher concentrations.

Nitrogen is so vital because it is a major component of chlorophyll, the compound by which plants use sunlight energy to produce sugars from water and carbon dioxide (i.e., photosynthesis). It is also a major component of amino acids, the building blocks of proteins. Without proteins, plants wither and die. Some proteins act as structural units in plant cells while others act as enzymes, making possible many of the biochemical reactions on which life is based. Nitrogen is a component of energy-transfer compounds, such as ATP (adenosine triphosphate). ATP allows cells to conserve and use the energy released in metabolism. Finally, nitrogen is a significant component of nucleic acids such as DNA, the genetic material that allows cells (and eventually whole plants) to grow and reproduce. Without nitrogen, there would be no life as we know it.

Structure of an Amino Acid

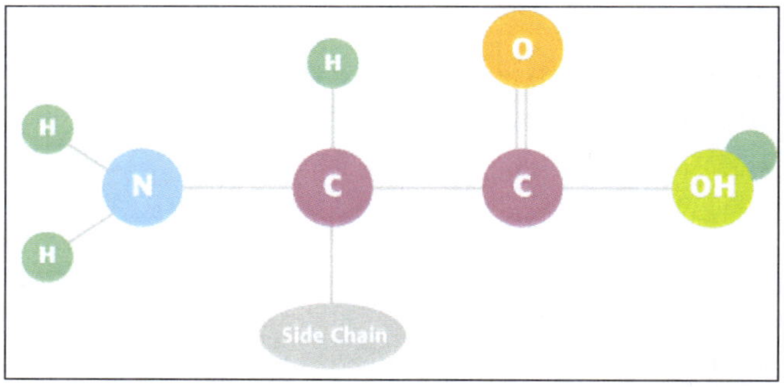

Nitrogen is essential for crops to achieve optimum yields. A critical component of amino acids in protein, it also increases protein content of plants directly.

Soil Nitrogen

Soil nitrogen exists in three general forms: organic nitrogen compounds, ammonium (NH_4^+) ions and nitrate (NO_3^-) ions.

At any given time, 95 to 99 percent of the potentially available nitrogen in the soil is in organic forms, either in plant and animal residues, in the relatively stable soil organic matter, or in living soil organisms, mainly microbes such as bacteria. This nitrogen is not directly available to plants, but some can be converted to available forms by microorganisms. A very small amount of organic nitrogen may exist in soluble organic compounds, such as urea, that may be slightly available to plants.

The majority of plant-available nitrogen is in the inorganic forms NH_4^+ and NO_3^- (sometimes called mineral nitrogen). Ammonium ions bind to the soil's negatively charged cation exchange complex (CEC) and behave much like other cations in the soil. Nitrate ions do not bind to the soil solids because they carry negative charges, but exist dissolved in the soil water, or precipitated as soluble salts under dry conditions.

Natural Sources of Soil Nitrogen

The nitrogen in soil that might eventually be used by plants has two sources: nitrogen- containing minerals and the vast storehouse of nitrogen in the atmosphere. The nitrogen in soil minerals is released as the mineral decomposes. This process is generally quite slow, and contributes only slightly to nitrogen nutrition on most soils. On soils containing large quantities of NH_4^+ rich clays (either naturally occurring or developed by fixation of NH_4^+ added as fertilizer), however, nitrogen supplied by the mineral fraction may be significant in some years.

Atmospheric nitrogen is a major source of nitrogen in soils. In the atmosphere, it exists in the very inert N_2 form and must be converted before it becomes useful in the soil. The quantity of nitrogen added to the soil in this manner is directly related to thunderstorm activity, but most areas probably receive no more than 20 lb nitrogen/acre per year from this source.

Bacteria such as Rhizobia that infect (nodulate) the roots of, and receive much food energy from, legume plants can fix much more nitrogen per year (some well over 100 lb nitrogen/acre). When the quantity of nitrogen fixed by Rhizobia exceeds that needed by the microbes themselves, it is released for use by the host legume plant. This is why well-nodulated legumes do not often respond to additions of nitrogen fertilizer. They are already receiving enough from the bacteria.

The Nitrogen Cycle

Nitrogen can go through many transformations in the soil. These transformations are often grouped into a system called the nitrogen cycle, which can be presented in varying degrees of complexity. The nitrogen cycle is appropriate for understanding nutrient and fertilizer management. Because microorganisms are responsible for most of these processes, they occur very slowly, if at all, when soil temperatures are below 50 °F, but their rates increase rapidly as soils become warmer.

The heart of the nitrogen cycle is the conversion of inorganic to organic nitrogen, and vice versa. As microorganisms grow, they remove H_4^+ and NO_3^- from the soil's inorganic, available nitrogen pool, converting it to organic nitrogen in a process called immobilization. When these organisms die and are decomposed by others, excess NH_4^+ can be released back to the inorganic pool in a process called mineralization. Nitrogen can also be mineralized when microorganisms decompose a material containing more nitrogen than they can use at one time, materials such as legume residues or manures. Immobilization and mineralization are conducted by most microorganisms, and are most rapid when soils are warm and moist, but not saturated with water. The quantity of inorganic nitrogen available for crop use often depends on the amount of mineralization occurring and the balance between mineralization and immobilization.

Ammonium ions (NH_4^+) not immobilized or taken up quickly by higher plants are usually converted rapidly to NO_3^- ions by a process called nitrification. This is a two-step process, during which bacteria called Nitrosomonas convert NH_4^+ to nitrite (NO_2^-), and then other bacteria, Nitrobacter, convert the NO_2^- to NO_3^-. This process requires a well-aerated soil and occurs rapidly enough that one usually finds mostly NO_3^- rather than NH_4^+ in soils during the growing season.

The nitrogen cycle contains several routes by which plant-available nitrogen can be lost from the soil. Nitrate-nitrogen is usually more subject to loss than is ammonium-nitrogen. Significant loss mechanisms include leaching, denitrification, volatilization and crop removal.

The nitrate form of nitrogen is so soluble that it leaches easily when excess water percolates through the soil. This can be a major loss mechanism in coarse-textured soils where water percolates freely, but is less of a problem in finer-textured, more impermeable soils, where percolation is very slow.

These latter soils tend to become saturated easily, and when microorganisms exhaust the free oxygen supply in the wet soil, some obtain it by decomposing NO_3^-. In this process, called denitrification, NO_3^- is converted to gaseous oxides of nitrogen or to N_2 gas, both unavailable to plants. Denitrification can cause major losses of nitrogen when soils are warm and remain saturated for more than a few days.

Losses of NH_4^+ nitrogen are less common and occur mainly by volatilization. Ammonium ions are basically anhydrous ammonia (NH_3) molecules with an extra hydrogen ion (H^+) attached. When this extra H^+ is removed from the NH_4 ion by another ion such as hydroxyl (OH^-), the resulting NH_3 molecule can evaporate, or volatilize from the soil. This mechanism is most important in high-pH soils that contain large quantities of OH^- ions.

Crop removal represents a loss because nitrogen in the harvested portions of the crop plant is removed from the field completely. The nitrogen in crop residues is recycled back into the system, and is better thought of as immobilized rather than removed. Much is eventually mineralized and may be reutilized by a crop.

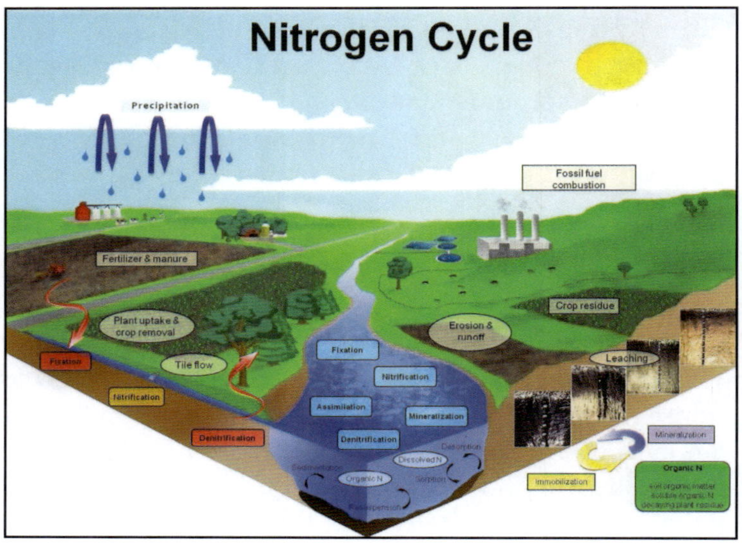

Plant Nitrogen needs and Uptake

Plants absorb nitrogen from the soil as both NH_4^+ and NO_3^- ions, but because nitrification is so pervasive in agricultural soils, most of the nitrogen is taken up as nitrate. Nitrate moves freely toward plant roots as they absorb water. Once inside the plant, NO_3^- is reduced to an NH_2 form and is assimilated to produce more complex compounds. Because plants require very large quantities of nitrogen, an extensive root system is essential to allowing unrestricted uptake. Plants with roots restricted by compaction may show signs of nitrogen deficiency even when adequate nitrogen is present in the soil.

Utilization of Nitrogen by Various Crops		
Crop	Yield per acre	Crop
Alfalfa	8 tons	432
Corn	180 bu	180
Soybeans	60 bu	294
Spring Wheat	80 bu	176
Winter Wheat	80 bu	152

Most plants take nitrogen from the soil continuously throughout their lives, and nitrogen demand usually increases as plant size increases. A plant supplied with adequate nitrogen grows rapidly and produces large amounts of succulent, green foliage. Providing adequate nitrogen allows an annual crop, such as corn, to grow to full maturity, rather than delaying it. A nitrogen-deficient plant is generally small and develops slowly because it lacks the nitrogen necessary to manufacture adequate structural and genetic materials. It is usually pale green or yellowish because it lacks adequate chlorophyll. Older leaves often become necrotic and die as the plant moves nitrogen from less important older tissues to more important younger ones.

On the other hand, some plants may grow so rapidly when supplied with excessive nitrogen that they develop protoplasm faster than they can build sufficient supporting material in cell walls. Such plants are often rather weak and may be prone to mechanical injury. Development of weak straw and lodging of small grains are an example of such an effect.

Fertilizer Management

Nitrogen Cycle

Nitrogen fertilizer rates are determined by the crop to be grown, yield goal and quantity of nitrogen that might be provided by the soil. Rates needed to achieve different yields with different crops vary by region, and such decisions are usually based on local recommendations and experience.

Factors that Determine the Quantity of Nitrogen Supplied by the Soil:

- The quantity of nitrogen released from the soil organic matter.
- The quantity of nitrogen released by decomposition of residues of the previous crop.
- Any nitrogen supplied by previous applications of organic waste.
- Any nitrogen carried over from previous fertilizer applications.

Such contributions can be determined by taking nitrogen credits (expressed in lb/acre) for these variables. For example, corn following alfalfa usually requires less additional nitrogen than corn following corn, and less nitrogen fertilizer is needed to reach a given yield goal when manure is applied. As with rates, credits are usually based on local conditions.

Soil testing is being suggested more often as an alternative to taking nitrogen credits. Testing soils for nitrogen has been a useful practice in the drier regions of the Great Plains for many years, and in that region, fertilizer rates are often adjusted to account for NO_3^- found in the soil prior to planting. In recent years, there has been some interest in testing cornfields for NO_3^- in the more humid regions of the eastern United States and Canada, utilizing samples taken in late spring, after crop emergence, rather than before planting. This strategy, the pre-side-dress nitrogen soil test (PSNT), has received a great deal of publicity and seems to provide some indication of whether additional side-dressed nitrogen is needed or not.

Fertilizer Placement

Placement decisions should maximize availability of nitrogen to crops and minimize potential losses. A plant's roots usually will not grow across the root zone of another plant, so nitrogen must be placed where all plants have direct access to it. Broadcast applications accomplish this objective. Banding does also when all crop rows are directly next to a band. For corn, banding anhydrous ammonia or urea ammonium nitrate (UAN) in alternate row middles is usually as effective as banding in each middle because all rows have access to the fertilizer.

Moist soil conditions are necessary for nutrient uptake. Placement below the soil surface can increase nitrogen availability under dry conditions because roots are more likely to find nitrogen in moist soil with such placement. Injecting side-dressed UAN may produce higher corn yields than surface application in years when dry weather follows side-dressing. In years when rainfall occurs shortly after application, subsurface placement is not as critical.

Subsurface placement is normally used to control nitrogen losses. Anhydrous ammonia must be placed and sealed below the surface to eliminate direct volatilization losses of the gaseous ammonia. Volatilization from urea and UAN solutions can be controlled by incorporation or injection. Incorporating urea materials (mechanically or by rainfall shortly after application) is especially important in no-till situations in which volatilization is aggravated by large amounts of organic material on the soil surface. Applying small amounts of "starter" nitrogen as UAN in herbicide sprays, however, is usually of little concern.

Placing nitrogen with phosphorus often increases phosphorus uptake, particularly when nitrogen is in the NH_4^+ form and the crop is growing in an alkaline soil. The reasons for the effect are not completely clear, but may be due to nitrogen increasing root activity and potential for phosphorus uptake, and nitrification of NH_4^+ providing acidity, which enhances phosphorus solubility.

Timing of Nutrient Application

Timing has a major effect on the efficiency of nitrogen management systems. Nitrogen should be applied to avoid periods of significant loss and to provide adequate nitrogen when the crop needs it most. Wheat takes up most of its nitrogen in the spring and early summer, and corn absorbs most nitrogen in midsummer, so ample availability at these times is critical. If losses are expected to be minimal, or can be effectively controlled, applications before or immediately after planting are effective for both crops. If significant losses, particularly those due to denitrification or leaching, are anticipated, split applications, in which much of the nitrogen is applied after crop emergence, can be effective in reducing losses. Fall applications for corn can be used on well-drained soils, particularly if the nitrogen is applied as anhydrous ammonia amended with N-Server; however, fall applications should be avoided on poorly drained soils, due to an almost unavoidable potential for significant denitrification losses. When most of a crop's nitrogen supply will be applied after significant crop growth or positioned away from the seed row (anhydrous ammonia or UAN banded in row middles), applying some nitrogen easily accessible to the seedling at planting ensures that the crop will not become nitrogen deficient before gaining access to the main supply of nitrogen.

Minimizing Fertilizer Losses

The major mechanisms for nitrogen fertilizer loss are denitrification, leaching and volatilization. Denitrification and leaching occur under very wet soil conditions, while volatilization is most common when soils are only moist and are drying.

Practices for Avoiding Nitrogen Fertilizer Losses

Using an NH_4^+ source of nitrogen acidifies the soil because the hydrogen ions (H^+) released during nitrification of the NH_4^+ are the major cause of acidity in soils. Over time, acidification and lowering of soil pH can become significant.

Acidity or Basicity of Different Nitrogen Sources			
Material	% Nitrogen	Approx. $CaCO_3$ Equivalent LB/ Ton of Material*	PER LB of N
Anhydrous Ammonia	82%	-2,960	1.80
Ammonia Sulfate	21%	-2,200	5.20
Urea	46%	-1,680	1.83
Diammonium Phosphate	18%	-1,400	3.8
Urea-Form	38%	-1,360	1.79
Monoammonium Phosphate	10%	-1,300	6.5
Ammonium Nitrate	33.5%	-1,180	116
Nitrogen Solutions	19-49%	-750 to -1,760	197-179
Calcium Nitrate	15%	+400	None
Potassium Nitrate	13%	+580	None
Sodium Nitrate	16%	+520	None

Nitrogen fertilizers containing NO_3^- but no NH_4^+ make the soil slightly less acidic over time, but are generally used in much lesser quantities than the others. The acidification due to NH_4^- nitrogen is a significant factor in the acidification of agricultural fields, but can easily be controlled by normal liming practices.

Fertilizing Legumes with Nitrogen

Because the Rhizobia bacteria that infect legume roots normally supply adequate nitrogen to the host plant, well-nodulated legumes rarely respond to additions of nitrogen fertilizer. Occasionally, however, soybeans may respond to applications of nitrogen late in the season, presumably because nitrogen fixation in the nodules has declined significantly. Such responses are quite erratic, though, and late-season applications of nitrogen to soybeans are not routinely recommended. The amount of atmospheric nitrogen fixed by non-symbiotic soil organisms varies with soil types, organic matter present and soil pH.

Approximate Amount of Nitrogen Fixed by Various Legume Crops	
Crop	LB/ACRE Nitrogen
Alfalfa	196
Ladino Clover	178
Sweet Clover	116
Red Clover	112
White Clover	103
Soybeans	98
Cowpeas	89

Lespedeza	85
Vetch	80
Garden Peas	71
Winter Peas	54
Peanuts	42

Additives for Nitrogen Fertilizers

N-Serve

N-Serve (nitrapyrin) is a proven material that selectively inhibits one of the bacteria responsible for nitrification. When added to an NH_4^+ nitrogen material, it delays its conversion to NO_3^- nitrogen for several weeks. It is most effective when mixed with anhydrous ammonia. This delayed nitrification protects the fertilizer from losses due to denitrification and leaching in seasons when excessive rainfall occurs during the period of inhibition. Using N-Serve is somewhat like buying an insurance policy that pays in years when problems develop.

AGROTAIN

AGROTAIN (NBPT) is a product that inhibits conversion of urea to ammonium carbonate, thereby reducing the potential for ammonia volatilization from urea materials, including UAN solutions. Like N-Server, it might be viewed as an insurance policy that will reduce potential nitrogen losses in seasons when cultivation or rain does not incorporate the urea into the soil soon after application. AGROTAIN is most useful when urea or UAN is applied without incorporation to the surface of fields with high levels of crop residue, such as in no-till situations, or fields with high pH levels at the surface.

Other Additives

Other nitrogen additives available on the market include ESN and Instinct. They can be used as part of the 4R Nutrient Stewardship strategy to keep nitrogen in its proper place at the time the plant needs it. Using nitrogen additive products is considered a best management practice for nitrogen management.

Phosphorus Fertilizers

Farmers and gardeners apply phosphate-based fertilizer to plants and soil to boost local phosphorus (P) levels. Young plants are especially vulnerable to deficient P levels, but almost every plant needs it to maintain healthy growth. Artificial phosphate fertilizer and manure or compost are used together to revitalize severely depleted soil. Phosphate fertilizers fall into one of four basic categories: rock phosphates, superphosphates, ammonium phosphates or polyphosphates. Each type of fertilizer is composed of different substances and chemical ratios:

- Phosphate fine 21% (Russia).
- Phosphate fine powder 24,5%(Russia).

Rock Phosphate (Apatite)

Deposits of rock phosphate form naturally in geological formations. It is harvested, ground to powder and added to soil as a source of phosphorus. There is a steady decline in the availability of rock phosphate on the market. The author cites the material's high transportation costs and poor water solubility as catalysts for its diminished use in agricultural practices. Rock phosphate still has limited use in domestic cultivation and is refined as an ingredient in other advanced fertilizers.

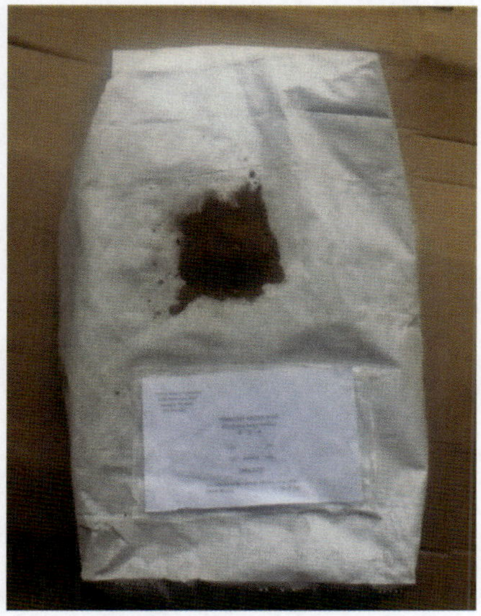

Rock Phosphates 36% P_2O_5 Togo & 38% P_2O_5 (Nauru origin) Rock Phosphates 24%, 26%, 28%, 29% and 30% P_2O_5 (Egyptian origin) Rock Phosphate 32% (Jordan origin).

Packing

- Bulk loaded on Bulk vessel.
- Bulk loaded inside 20-foot containers.
- 1000 kg bags, 50 kg bags.

Single Super Phosphate

Normal super phosphate is a stronger and more soluble compound than rock phosphate, but it is significantly less potent than concentrated super phosphate or ammonium compounds. The newer concentrated compound contains 46% phosphate, while the normal variety has 20%. Neither of the two supplies nitrogen, a desirable nutrient found in other fertilizers.

Superphosphate

- Appearance: Granulated product without mechanical admixtures.
- Application: Used in agriculture as a mineral fertilizer.
- Assimilable phosphates by wt converted to P_2O_5, % 19 ± 1.

Granular Triple Super Phosphate

Triple super phosphate is made by the action of phosphoric acid on raw rock phosphate (also called appatite). It typically contains 46% phosphate, soluble in neutral ammonium citrate and water. Of this, typically 90% is soluble in water alone. Because the phosphate on TSP is never completely water-soluble, it cannot be used in the manufacture of clear-liquid fertilizers.

MAP (Monoammonium Phosphate)

N:P:K = 12-61-0 MAP application:

- As a high grade fertilizer for the production of complex fertilizers as well, could be used itself as solution: additional fertilizers, in hotbed for vegetables and flowers.
- As feed admixtures for the stock-breeding.
- As additional fertilizers for yeast production (feed, alcohol, bakery, beer), medicines; - For textile impregnation, plywood and timber for fire barrier.
- For preparing fire barrier solutions used in wooden construction treatment.
- For the production of fire-extinguisher powder.

DAP (Diammonium Phosphate) N:P:K = 21-53-0;18-46-0

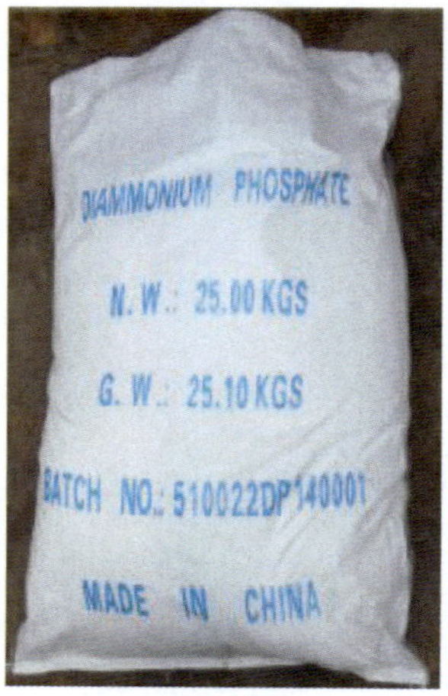

Diammonium phosphate fertilizer is a plant food that is rich in the mineral phosphorus. It is widely used among farmers in the production of grain and row crops, such as corn, wheat, rye and some lawn grasses. Some farmers may choose to blend this type of plant food with other fertilizer forms to meet the specific needs of their crops and soils. It may be applied using a variety of methods before, during, or after crops have been planted.

How Commercial Phosphate Fertilizer is Manufactured

Materials

Rock phosphate is the raw material used to manufacture most commercial phosphate fertilizers on the market. In the past, ground rock phosphate itself has been used as a source of P for acid soils. However, very little rock phosphate is currently used in agriculture due to low availability of P in this native material, high transportation costs and small crop responses.

Process

Most commercial phosphate fertilizer manufacturing begins by producing phosphoric acid.

The generalized diagram in figure below shows the steps taken to manufacture various phosphate fertilizers. Phosphoric acid is produced by either a dry or wet process.

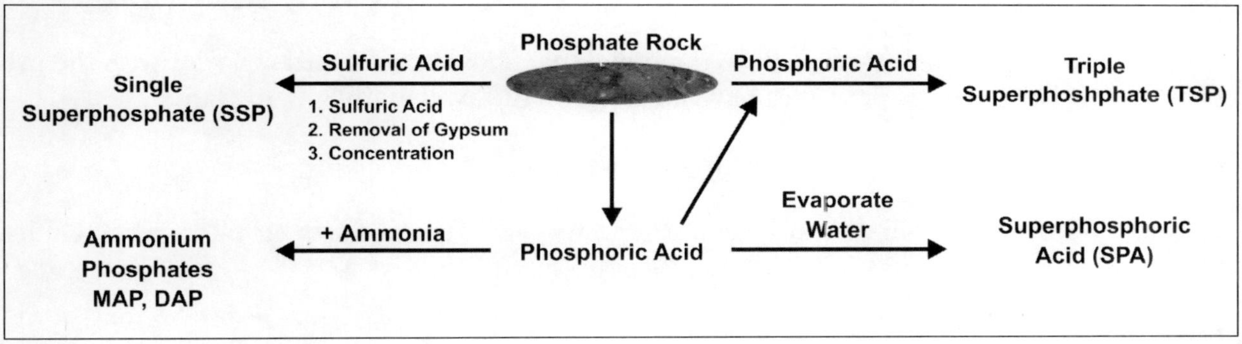

Dry vs. Wet Process

In the dry process, an electric furnace treats rock phosphate. This treatment produces a very pure and more expensive phosphoric acid – frequently called white or furnace acid – primarily used in the food and chemical industry.

Fertilizers that use white phosphoric acid as the P source are generally more expensive because of the costly treatment process. The wet process involves treating the rock phosphate with acid-producing phosphoric acid – also called green or black acid – and gypsum, which is removed as a by-product. The impurities that give the acid its color haven't been a problem in the production of dry fertilizers.

Orthophosphoric Acid

Both the wet and dry treatment processes produce orthophosphoric acid, the phosphate form that's taken up by plants.

The phosphoric acid produced by either the wet or dry process is frequently heated, driving off water and producing a superphosphoric acid. The phosphate concentration in superphosphoric acid usually varies from 72 to 76 percent.

The P in this acid is present as both orthophosphate and polyphosphate. Polyphosphates consist of a series of orthophosphates that have been chemically joined together. Upon contact with soils, polyphosphates revert back to orthophosphates.

Adding Ammonia

Ammonia can be added to the superphosphoric acid to create liquid or dry materials containing both nitrogen (N) and P. The liquid, 10-34-0, is the most common product.

The 10-34-0 can be mixed with finely ground potash (0-0-62), water and urea-ammonium nitrate solution (28-0-0) to form 7-21-7 and related grades. The P in these products is present in both the orthophosphate and polyphosphate form.

When ammonia is added to phosphoric acid that hasn't been heated, it produces monoammonium phosphate (11-52-0) or diammonium phosphate (18-46-0), depending on the ratio of the mixture. The P present in these two fertilizers is in the orthophosphate form.

Cost and Outcome

The cost of converting rock phosphate to the individual phosphate fertilizers varies with the process. More importantly, the processes have no effect on the availability of P to plants.

Phosphate Fertilizer Terminology and Sources

Selecting a phosphate fertilizer can be confusing due to all the products on the market. Understanding the terminology may help avoid some of the confusion.

Water – Soluble

Fertilizer samples analyzed by a control laboratory are placed in water, then the percentage of the total phosphate that dissolves is measured. This percentage is referred to as water-soluble phosphate.

Citrate – Soluble

The fertilizer material that isn't dissolved in water is then placed in an ammonium citrate solution. The amount of P dissolved in this solution is measured and expressed as a percentage of the total in the fertilizer material.

Phosphate measured with this analytical procedure is referred to as citrate-soluble.

Available

The sum of the water-soluble and citrate-soluble phosphates is considered to be the percentage that's available to plants and is the amount guaranteed on the fertilizer label. Usually, the citrate-soluble component is less than the water-soluble component.

Comparison Chart: Common Fertilizer Sources

Table: Percentages of water-soluble and available phosphate in several common fertilizer source.

P_2O_5 source	N	Total	AvailableP_2O_5	Water soluble* P_2O_5
Superphosphate (OSP)	0%	21%	20%	85%
Concentrated Superphosphate (CSP)	0%	45%	45%	85%
Monoammonium Phosphate (MAP)	11%	49%	48%	82%
Diammonium Phosphate (DAP)	18%	47%	46%	90%
Ammonium Polyphosphate (APP)	10%	34%	34%	100%
Rock Phosphate	0%	34%	38%	0%
*Water-soluble data are a percent of the total P_2O_5.				

Organic Phosphorus Sources

Organic P fertilizers have been used for centuries as the P source for crops. Even with the advent of P fertilizer technology processes, organic P sources from animal manures – including composts – and sewage sludge are still very important.

Organic vs. Inorganic

From a fertilizer/nutrient management perspective, the major differentiating factor is the availability of P.

As with any fertilizer products, especially those with varying analysis, do a chemical analysis. Then use an availability coefficient to determine the available P as a portion of the reported total P.

Phosphorus from manure or sludge should be comparable to P from inorganic fertilizer. So, if a producer has a P recommendation for 30 pounds per acre of P₂O₅, applying approximately 65 pounds of 18-46-0 (DAP) or 6 tons of 11-6-9 (manure; 80 percent available P coefficient) should provide equivalent results.

Chemical Composition

The P contained in organic P sources combines inorganic and organic P. Essentially, all inorganic P is in the orthophosphate form, which is the form taken up by growing plants.

Chemical Makeup

Diet fed to the animal has some control over this chemical makeup. Consider P feed supplements and the fact that many could be considered P fertilizers as well. Generally, 45 to 70 percent of manure-P is inorganic P. Organic P constitutes the remaining total P.

Decomposition

Much of the organic P is easily decomposable in the soil, but factors such as temperature, soil moisture and soil pH all have a bearing on the P mineralization rate. The final decomposition product is orthophosphate P compounds.

Available P

The combination of the organic-inorganic P ratios in the organic P sources and the soil environment affect the availability coefficient for organic P. Most animal manure research interpretations indicate that approximately 60 to 80 percent of the total P is available to crops in the first year.

Due to the chemical composition of other organic P sources such as bone meal, expect lesser amounts of plant-available P compared to total P.

How Crops Respond to Phosphate Fertilizers

If the level of available P in the soil isn't adequate for optimum crop growth, use phosphate fertilizers to ensure adequate amounts of this nutrient in the solution phase.

Corn Yield

Optimal Soil Test Levels

Numerous research projects have demonstrated that agronomic crops will respond to phosphate fertilization if soil test levels are in the very low, low and medium ranges, or below 15 parts per million (ppm) in the Bray-1 test or 11 ppm in the Olsen test.

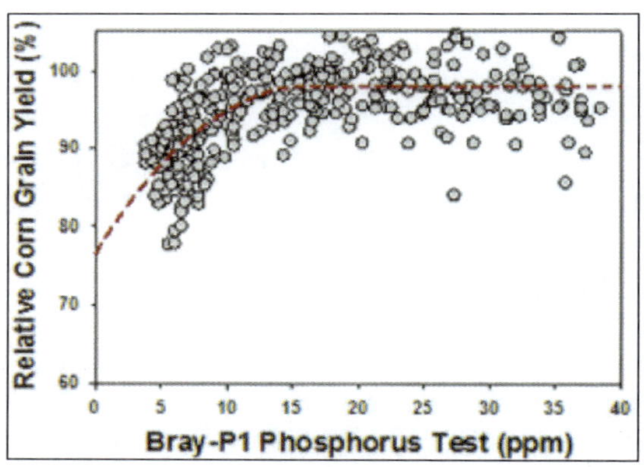

Crop Removal

Crop removal is common in many areas of the state. In the example in table, banding the P at a lower rate resulted in the same yield as those based on crop removal recommendations. This illustrates the effect that banding P can have on reducing the corn crop's overall P requirements.

Application Method

Shows the effect of banded starter (two inches beside and below the seed) and broadcast phosphate on corn yield when soil test levels for phosphorus are medium. Data are an average of two years of data collected at the West Central Research and Outreach Center in Morris.

The ability of the banded fertilizer application to supply a crop's entire P requirement can depend on the type of band used and the soil test. Banding liquid fertilizer on the seed is common for corn and sugarbeet.

When banding on the seed, use a low rate. This is because there's potential to reduce emergence due to high salts or ammonia forming near the seed.

The example in figure shows that a small rate of phosphate banded with the seed can provide maximum yield for corn with medium soil test P levels. However, it's not enough to maximize yield with low soil test P levels.

In contrast, recent data has shown that a small rate of fertilizer banded with the seed is better than higher rates of broadcast P for sugarbeet.

Table: How banded starter and broadcast phosphate affect corn yield.

P_2O_5 rate	Placement	Recommendation basis	Grain yield
0	Control	-	178 bushels per acre
49	Broadcast	One-year crop removal	187 bushels per acre
85	Broadcast	Two-year crop removal (C-Sb)	185 bushels per acre
35	Broadcast	University of Minnesota	180 bushels per acre
25	Starter	University of Minnesota	188 bushels per acre

The ability of the banded fertilizer application to supply a crop's entire P requirement can depend on the type of band used and the soil test. Banding liquid fertilizer on the seed is common for corn and sugarbeet.

When banding on the seed, use a low rate. This is because there's potential to reduce emergence due to high salts or ammonia forming near the seed.

The example in figure below shows that a small rate of phosphate banded with the seed can provide maximum yield for corn with medium soil test P levels. However, it's not enough to maximize yield with low soil test P levels.

In contrast, recent data has shown that a small rate of fertilizer banded with the seed is better than higher rates of broadcast P for sugarbeet.

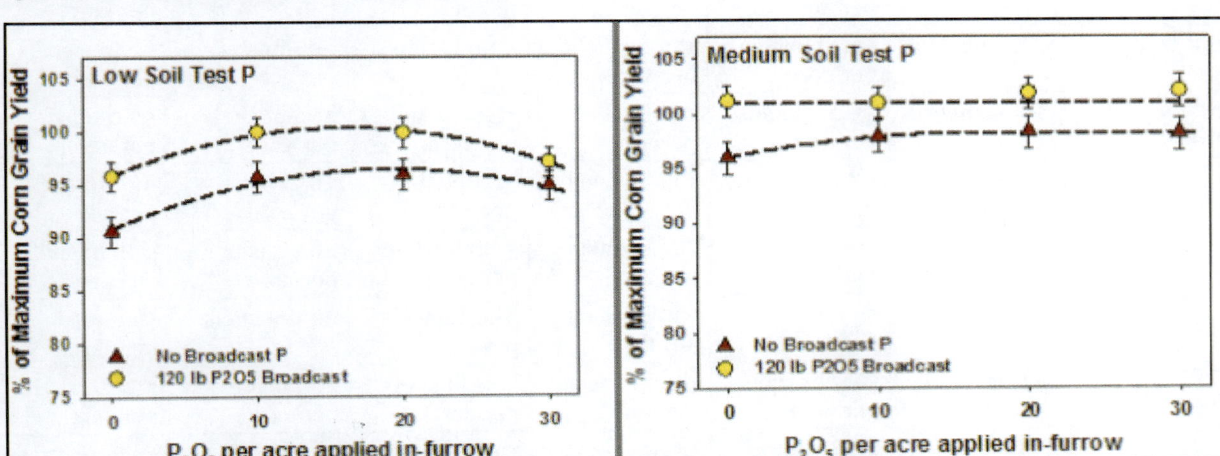

How corn responds to a combination of broadcast phosphorus and in-furrow (on seed) banded phosphorus as 10-34-0

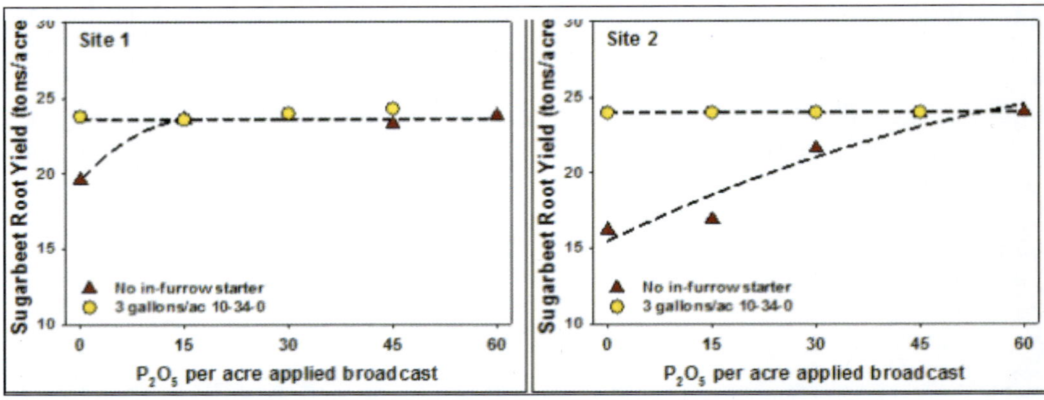

How sugarbeet responds to broadcast phosphorus and three gallons per acre in-furrow (on seed) banded phosphorus as 10-34-0.

Corn and Soybean Yield

Soil Test Values

This example illustrates the effect that starting soil test level, soil type and crop can have on the response to P.

Corn grain yield responded to P at two of the locations, Lamberton and Morris, while soybean only responded at Morris, which had the lowest starting soil test value for P. Understanding which crops respond better at which soil test values is important to ensure maximum return on investment when applying P.

Application Method

How the rate of phosphate broadcast applied affects corn and soybean yield.

Phosphorus rate	Lamberton: Corn	Lamberton: Soybean	Morris: Corn	Morris: Soybean	Saint Charles: Corn	Saint Charles: Soybean
0 P_2O_5 per acre	169 bushels per acre	52.2 bushels per acre	175 bushels per acre	48.0 bushels per acre	201 bushels per acre	51.1 bushels per acre
20 P_2O_5 per acre	175 bushels per acre	53.7 bushels per acre	183 bushels per acre	50.6 bushels per acre	208 bushels per acre	52.9 bushels per acre
40P_2O_5 per acre	178 bushels per acre	53.0 bushels per acre	189 bushels per acre	51.0 bushels per acre	206 bushels per acre	52.0 bushels per acre
80 P_2O_5 per acre	179 bushels per acre	52.7 bushels per acre	193 bushels per acre	51.4 bushels per acre	205 bushels per acre	51.6 bushels per acre
120 P_2O_5per acre	178 bushels per acre	53.0 bushels per acre	193 bushels per acre	52.8 bushels per acre	202 bushels per acre	51.2 bushels per acre
160 P_2O_5per acre	181 bushels per acre	53.6 bushels per acre	190 bushels per acre	51.0 bushels per acre	203 bushels per acre	52.2 bushels per acre
240 P_2O_5per acre	184 bushels per acre	54.5 bushels per acre	196 bushels per acre	52.1 bushels per acre	203 bushels per acre	51.4 bushels per acre
–	22 Bray-P1 soil test P (ppm)	22 Bray-P1 soil test P (ppm)	14 Bray-P1 soil test P (ppm)	10 Bray-P1 soil test P (ppm)	14 Bray-P1 soil test P (ppm)	14 Bray-P1 soil test P (ppm)

Response of Other Crops

Crop response to P application varies:

- Alfalfa: Will respond to levels up to 40 ppm (soil test Bray P-1 value).
- Wheat and soybean: Will only respond up to 10 to 15 ppm (soil test Bray P-1 value).
- Corn: Will respond to levels up to 15 to 20 ppm.
- Potato: Will respond to levels above 30 ppm. However, response is more likely when soil test P is below 30 ppm.

Predicting the Need for Phosphate Fertilizer

Phosphorus soil tests measure soil's ability to supply P to the soil solution for plant use, but do not measure the total quantity of available P. These tests provide an availability index of P in soils that relates to the phosphate fertilizer's ability to provide an economically optimal increase in yield.

The relationship between the P determined by a soil test and the phosphate fertilizer requirements are developed from the results of numerous research trials that measured various rates of applied phosphate and yields.

Corn in Minnesota

Table summarizes recent data on corn response to P in Minnesota. For various starting soil test values, table gives:

- The percentage of times that applying P resulted in a measurable increase in corn yield.
- The average yield achieved when no P was applied.

It's important to note that there's always a possibility that applying P will increase the crop's yield. As shown in table, applying P in the high and very high categories increased corn grain yield 14 and 9 percent of the time, respectively.

However, the average yield produced in those categories was within 1 percent of the maximum of maximum. Maintaining high to very high soil test levels will ensure maximum yield potential, but the low probability of response to P will result in a poor economic return from high rates of applied P.

Table: Corn grain yield response to applied P fertilizer based on soil test category.

Bray P-1 or Olsen soil test P category	Expected time P fertilizer will increase corn grain yield	Expected yield without P fertilizer
Very low	87%	87%
Low	83%	90%
Medium	27%	98%
High	13%	99%
Very high	7%	99%

Measuring Soil P Status

Olsen and Bray-1 Procedures

Two laboratory procedures are used to measure the P status of Minnesota soils:

- Olsen procedure: Use when the soil pH is 7.4 or greater.
- Bray-1 procedure: Use when the soil pH is less than 7.4.

Both soil tests have been correlated and calibrated with yield response. The phosphate recommendations in Minnesota are based on those correlation values.

Some soil testing laboratories analyze soils with both a weak Bray (P-1) and a strong Bray (P-2) procedure. Bray P-2 results have not been correlated and calibrated to the crop response to phosphate fertilizer in Minnesota and aren't useful in predicting the amount of phosphate fertilizer to apply.

There are several situations where the soil pH is greater than 7.4 and the P value from the Bray-1 procedure is greater than the P value from the Olsen procedure.

When soil samples are analyzed by both the Olsen and Bray-1 procedures, research data indicates that phosphate fertilizer recommendations should be based on the greater value. You can also use plant analysis as an aid in determining the availability of P in soils.

Mehlich-3 Soil Test

Several states in the Corn Belt use the Mehlich-3 soil test, but it's not recommended in Minnesota.

The Mehlich-3 soil test will typically result in soil P test levels 0 to 5 percent greater than the Bray-P1 test when soil pH is 7.5 or less. The Mehlich-3 test has been found to be less reliable for soils with excess carbonates and a pH greater than 7.5.

Symptoms of P Deficiency

Symptoms of P deficiency aren't obvious or easily identifiable for most crops in Minnesota. For most crops, a shortage of P reduces plant size.

Less Plant Growth

Figure shows less plant growth due to a shortage of P in potatoes. This lack of growth is typical for crops such as potato and soybean when P is deficient.

Purpling

For corn, a severe P deficiency inhibits the translocation of carbohydrates within the plant. This leads to a purple color on the margins of the leaves.

The purpling is usually most evident in young corn plants because there's a greater demand for P early in the growing season. Figure 6 shows a P-deficient corn plant.

Some hybrids have a purple appearance early in the growing season regardless of the P supply in the soil. This can be a genetic response to stress caused by cold temperatures. Don't confuse this hybrid characteristic with P deficiency.

Reduced plant growth in potato due to a phosphorus deficiency

Purpling on the edge of corn leaves due to phosphorus deficiency

P Sufficiency Level by Crop Type

When to use Plant Analysis as a Management Tool

It's important to relate the interpretation of the analytical results to the growth stage. The concentration of P in plant tissue usually decreases as the plant matures. Table summarizes some interpretations of P concentrations for several crops.

Table: Sufficiency levels of phosphorus for major agronomic crops, vegetables and fruits grown in Minnesota.

Crop	Plant part	Time	Sufficiency range
Alfalfa	Tops (six inches new growth)	Prior to flowering	0.26 to 0.70%
Apple	Leaf from middle of current terminal shoot	July 15 to Aug. 15	0.09 to 0.40%
Blueberry	Young mature leaf	First week of harvest	0.10 to 0.40%
Broccoli	Young mature leaf	Heading	0.30 to 0.70%
Cabbage	Half-grown young wrapper leaf	Heads	0.33 to 0.75%
Carrot	Young mature leaf	Mid-growth	0.20 to 0.30%
Cauliflower	Young mature leaf	Buttoning	0.33 to 0.80%
Edible bean	Most recently matured trifoliate	Bloom stage	0.30 to 0.50%
Field corn	Whole tops	Less than 12 inches tall	0.25 to 0.50%
Field corn	Base of ear	Initial silk	0.30 to 0.60%
Grape	Petiole from young mature leaf	Flowering	0.20 to 0.46%
Pea	Recently mature leaflet	First bloom	0.30 to 0.80%
Potato	Fourth leaf from tip	40 to 50 days after emergence	0.25 to 0.50%
Potato	Petiole from fourth leaf to tip	40 to 50 days after emergence	0.22 to 0.40%
Raspberry	Leaf 18 inches from tip	First week in August	0.20 to 0.50%

Soybean	Trifoliate leaves	Early flowering	0.30 to 0.60%
Spring wheat	Whole tops	As head emerges from boost	0.20 to 0.50%
Strawberry	Young mature leaf	Mid-August	0.20 to 0.35%
Sweet corn	Ear leaf	Tasseling to silk	0.25 to 0.40%
Sugar beet	Recently matured leaves	50 to 80 days after planting	0.45 to 1.10%

Managing Phosphate Fertilizers

Because P isn't mobile in soils, placing phosphate fertilizers is a major management decision in crop production systems. There's no special placement that's ideal for all crops. Decisions about placing phosphate fertilizers are primarily affected by the intended crop and P soil test level.

Corn

For corn and small grain production, the needed phosphate fertilizer can be:

- Broadcast and incorporated before planting.
- Applied in a band away from the seed row as a starter fertilizer at planting.
- Directly on the seed at planting, if small amounts are needed.

With small grains, you can apply the amount of needed phosphate with a drill or air seeder at planting. Corn starter fertilizer is usually separated from the seed by approximately 1 inch of soil.

The banded application is a very efficient way to use phosphate fertilizer, as you can cut the recommended broadcast application rates in half. Results suggest you can place a small amount of fertilizer directly on the corn seed with the planter. However, the rate applied may not satisfy the amount needed for corn if soil test phosphorus is low.

Soybean

Soybean research trials have shown that greater grain yields are produced if the needed phosphate is broadcast and incorporated before planting, compared to a band application. This response is the opposite of corn and small grain, and may best be explained by differences in the development of the respective root systems.

Sugar Beet

For sugar beet, current research suggests that seed row placement of 15 pounds of phosphate will produce similar yields as 45 to 60 pounds of phosphate broadcast to the soil.

Other Crops

For other row crops, there isn't enough research to suggest a preferred method of phosphate placement. Applying phosphate for alfalfa and other forage crops is more efficient when done before stand establishment, when the fertilizer can be incorporated prior to seeding.

Grasses and legumes develop a large number of small roots near the soil surface. This means these crops can absorb phosphate fertilizers that are annually broadcast to established stands, if additional fertilizer is required.

Potassium Fertilizers

Potassium is essential in nearly all processes needed to sustain plant growth and reproduction. Plants deficient in potassium are less resistant to drought, excess water, and high and low temperatures. They are also less resistant to pests, diseases and nematode attacks. Because potassium improves the overall health of growing plants and helps them fight against disease, it is known as the "quality" nutrient. Potassium affects quality factors such as size, shape, color and vigor of the seed or grain, and improves the fiber quality of cotton.

Potassium increases crop yields because it:

- Increases root growth and improves drought tolerance.
- Builds cellulose and reduces lodging.
- Activates at least 60 enzymes involved in growth.
- Aids in photosynthesis and food formation.
- Helps translocate sugars and starches.
- Produces grains rich in starch.
- Increases protein content of plants.
- Maintains turgor, reduces water loss and wilting.
- Helps retard the spread of crop diseases and nematodes.

Potassium Uptake by Crops

Time of potassium uptake varies with different plants. However, plants generally absorb the majority of their potassium at an earlier growth stage than they do nitrogen and phosphorus.

Experiments on potassium uptake by corn showed that 70 to 80 percent was absorbed by silking time, and 100 percent was absorbed three to four weeks after silking. Translocation of potassium from the leaves and stems to the grain was much less than for phosphorus and nitrogen. The period during grain formation is apparently not a critical one for supply of potassium.

Crop	Yield	Uptake (K_2O)
Alfalfa	10 lb/ton	600 lb/acre
Corn	200 bu/acre	266 lb/acre
Cotton (Lint)	1,500 lb/acre	210 lb/acre

Grain Sorghum	800 lb/acre	240 lb/acre
Peanuts	4,000 lb/acre	185 lb/acre
Soybeans	60 bu/acre	205 lb/acre
Wheat	80 bu/acre	162 lb/acre

Potassium Removal by Crops

Nutrient uptake or utilization is an important consideration, but crops take up far more potassium than they remove with the harvested portion. For example, a 200 bu/acre corn crop takes up or utilizes about 266 lb/acre of potash. But when the corn is harvested as grain, only 0.25 lb/bu is removed, or 50 lb/acre K_2O is harvested and removed from the field. However, if the crop were harvested as silage, then 7.3 lb/ton K_2O are vested and removed from the field. Therefore, a 32-ton/acre silage crop would remove 234 lb/acre K_2O. Harvest management is the major consideration in developing a potash fertilization program. Crops harvested in which the whole plant is removed from the field, like alfalfa hay, must have more potash applied than crops where only grain, lint or fruit are removed. Often with hay and silage crops, removal is an excellent guide for planning the potash fertilization program. With other crops, such as grain, soil tests offer the best guide.

Crop	Removal (K_2O)
Alfalfa	49.0 lb/ton
Corn Grain	0.25 lb/bu
Corn Silage	7.30 lb/ton
Coflon (Lint)	19.0 lb/bale
Grain Sorghum	0.27 lb/bu
Peanuts (Nuts)	17.0 lb/ton
Rice Grain	0.16 lb/bu
Soybean Grain	1.18 lb/bu
Sugarcane	3.50 lb/ton
Fescue (DM)	54.0 lb/ton
Spring Wheat Grain	0.33 lb/bu
Winter Wheat Grain	0.29 lb/bu

Potassium Deficiency Symptoms

Potassium is a highly mobile element in the plant and is translocated from the older to younger tissue. Consequently, potassium deficiency symptoms usually occur first on the lower leaves of the plant, and progress toward the top as the severity of the deficiency increases. One of the most common signs of potassium deficiency is the yellow scorching, or firing (chlorosis), along the leaf margin. In severe cases, the fired margin of the leaf may fall out. However, with broadleaf crops, such as soybeans and cotton, the entire leaf may shed, resulting in premature defoliation of the crop.

Potassium-deficient crops grow slowly and have poorly developed root systems. Stalks are weak, and lodging of cereal crops such as corn and small grain is common. Legumes are not strong competitors for soil potassium and are often crowded out by grasses in a grass-legume pasture. When potassium is not sufficient, winter killing of perennial crops such as alfalfa and grasses can occur.

Symptoms in Corn

Firing or scorching appears on outer edge of leaf while midrib remains green. May be some yellow striping on lower leaves. (Sorghum and most grasses also react this way.) Poor root development, defective nodal tissues, unfilled, chaffy ears, and stalk lodging are other symptoms in corn.

Potassium in Soil

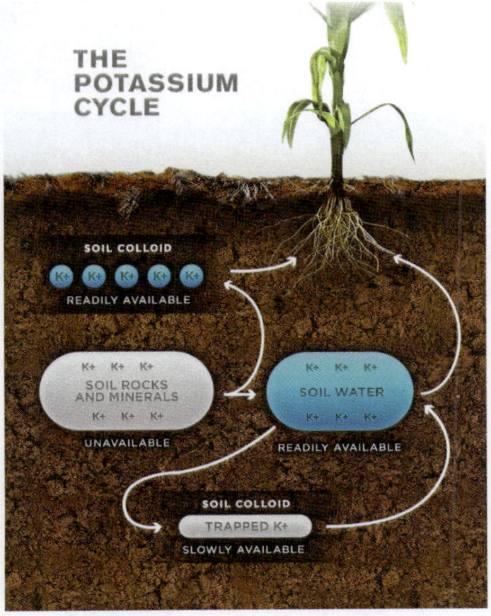

Relatively Unavailable Potassium

From 90 to 98 percent of the total potassium present in soils is found in insoluble primary minerals that are resistant to chemical breakdown. They release potassium slowly, but in small quantities compared to total needs of growing crops.

Slowly Available Potassium

This form makes up 1–10 percent of the total potassium supply, and may originate from dissolved primary minerals or from potassium fertilizers. This potassium is attracted to the surface of clay minerals, where it may be firmly bound or fixed between the clay layers in a form slowly available to plants. The actual amount available depends on the type and amount of clay present.

Readily Available Potassium

Readily available forms of potassium make up only 0.1 to 2 percent of the total potassium in the soil, and consist of potassium dissolved in the soil solution and held on the exchange positions of the clay and organic matter. This potassium is "exchangeable" because it can be replaced by other positively charged ions (cations) such as hydrogen, calcium and magnesium. This exchange happens rapidly and frequently. The potassium in the soil solution may be taken up by the plant or lost from the soil by leaching, especially on sandy, coarse-textured soils.

Potassium and Balanced Crop Nutrition

Adequate supplies of other plant nutrients are required to obtain maximum responses to potassium fertilizer; however, there are several unique relations between potassium and other nutrients.

High-potassium fertilization can decrease the availability of magnesium to the plant, and may result in magnesium deficiency of crops grown on soils that are already low in magnesium. This problem is often encountered with crops grown on sandy soils, particularly in the coastal plain soils of the southern United States. Conversely, crops grown on soils high in magnesium can suffer potassium deficiency, especially if the soils are high phosphorus and low in potassium. This problem is especially severe in the soils of the Mississippi River flood plain.

Leaching of potassium in acidic, sandy soils may be reduced by liming the soil to a pH of 6.2 to 6.5; however, applications of high rates of limestone to a soil low in potassium may induce potassium deficiency of crops growing on those soils. This problem occurs more on soils with predominantly 2:1 type clays (such as montmorillonite clays) rather than the 1:1 type (such as kaolinitic clays).

Potash

Fertilizer potassium is sometimes called "potash", a term that comes from an early production technique where potassium was leached from wood ashes and concentrated by evaporating the leachate in large iron pots ("pot-ash"). Clearly, this practice is no longer practical and is not environmentally sustainable. In food production, potassium is removed from the soil in harvested crops and must be replaced in order to maintain future crop growth.

Over 350 million years ago, the huge Devonian Sea was slowly drying up in the area of Central Canada and northern U.S., leaving behind concentrated salts and minerals. This process continues today in places such as the Great Salt Lake and the Dead Sea.

These ancient marine salts are now recovered and used in a variety of useful ways, with the majority being used as potassium fertilizer. Potassium is a natural plant food because fertilizers such as potassium chloride and potassium sulfate are widely found in nature. Fortunately, there are huge reserves of potash in the earth that can meet our need for this nutrient for many centuries to come. This fertilizer is clearly not an artificial or manufactured chemical, since it comes directly from the earth and is simply recycled through very long geological processes.

Potassium is an important mineral required for human health. Since potassium is not stored in the body, it is necessary to continually replace this nutrient on a regular basis with potassium-rich foods. Diets high in potassium and low in sodium have been shown to be beneficial for avoiding high blood pressure.

Potassium is essential for plant health and there must be an adequate supply in the soil to maintain good growth. When the potassium supply is limited, plants have reduced yields, poor quality, utilize water less efficiently, and are more susceptible to pest and disease damage.

In many parts of the world, agricultural soils are gradually becoming depleted of potash. Some soils were high in potassium when they were first cultivated long ago. However, after many years of intensive cropping and repeated nutrient removal during harvest, many fields now require regular inputs of potash to maintain their productivity.

High yielding crops remove large amounts of potassium in the harvested portion of the crop. For example, harvesting 9 ton alfalfa/A will remove over 450 lb K_2O. Similarly, a potato yield of 450 cwt/A removes 500 lb K_2O, and harvesting 40 ton/A of tomatoes will take off over 450 lb K_2O/A. But these high rates of nutrient removal are not usually being matched with fertilization. For example, In Idaho an average of four pounds of potash are removed in crops for every pound that is added back. In the Pacific coast states, over two pounds of potash are removed on average for every pound returned to the field as fertilizer. It's little wonder that K deficiency is becoming a more common occurrence in agricultural fields.

There are many excellent sources of potassium that can be used to sustain a productive and healthy ecosystem and replenish the soil's nutrient reserve. So which one should you use? Some of the most popular include:

- Potassium chloride (Muriate of potash) (KCl; 0-0-60).
- Potassium sulfate (Sulfate of potash)(K_2SO_4; 0-0-50-18S).
- Potassium-magnesium sulfate(K_2SO_4-$2MgSO_4$; 0-0-22-22S-11 Mg).
- Potassium thiosulfate($K_2S_2O_3$; 0-0-25-17S).
- Potassium nitrate(KNO_3; 13-0-44).

The potassium in all these fertilizers is identical and this nutrient will be rapidly available to the plant regardless of the source. The primary difference is in the companion nutrients that come along with the potassium.

- Potassium Chloride: The importance of chloride is frequently overlooked, but it is an essential nutrient for plant growth. Recent research has demonstrated that many crops respond favorably to chloride applications with greater yield and quality. Like any soluble fertilizer, salt-induced damage can result if large amounts are placed in close proximity to seeds or seedlings. Potassium chloride is usually the least expensive source of potash.
- Potassium Sulfate: All crops require an adequate supply of sulfur to develop proteins and enzymes. Sulfur-deficient plants appear light green and have reduced yields and quality. Sulfate that is present in potash fertilizers is immediately available for plant uptake.
- Potassium-Magnesium Sulfate: Because its vital role in chlorophyll, magnesium deficiency is first exhibited by yellow leaves in the lower part of the plant. Magnesium requirements vary considerably, with legumes generally containing more of this element than grasses.

- Potassium Nitrate: An abundant supply of nitrogen is essential for all high-yielding crops. For crops that prefer a nitrate source to an ammonium source of nitrogen, this potash source can be a good option.

There are many excellent potash sources available for meeting the nutrient requirements of crops. When making a decision on which source to use, choose the one that meets your needs and provides the accompanying anion that will help keep your high-yielding crops in top shape.

Fertilizer Raw Materials

Most fertilizer blends consist largely of nitrogen (N), phosphorus (P) and potassium (K). These three elements make up the largest bulk and cost of most blends. Nitrogen, phosphorous and potassium (N-P-K) are also present in plant tissue on a percentage scale on a dry weight basis. The Fertilizer Institute has an abundance of online resources available that detail the origin and use of fertilizer components.

- Nitrogen (N): Available worldwide, nitrogen gas is a major component of the atmosphere (78 percent). Based on this fact, one would expect that nitrogen itself would be almost free for plants, much like the hydrogen, oxygen and carbon that plants take directly from the air or water. The process of manufacturing nitrogen-based fertilizer combines naturally occurring nitrogen from the atmosphere with natural gas to form anhydrous ammonia. This substance can then be applied directly to crops as a nitrogen fertilizer or used to make other fertilizer products including: urea and ammonium nitrate. While it may sound simple, this process requires a significant amount of energy that not everyone can produce or afford.
- Phosphorus (P): Phosphorous is found in slightly less than 1 percent of the earth's crust. Phosphorous-based fertilizer comes from the phosphates originating in rock deposits as fossilized remains of ancient marine life or from volcanic activity. Mined rock phosphate from these natural geological deposits is combined with other chemicals to produce triple super-phosphate or mono-ammonium phosphate (MAP). China, the United States, India, Canada and Brazil have the greatest phosphorous resources. Although many countries have found reserves of phosphate rock, the energy-intensive extraction and processing steps make it feasible for only a few countries to supply phosphate-based fertilizer.
- Potassium (K): Potassium is the seventh most common element in the earth's crust comprising approximately 2.5 percent. Potassium-based fertilizer is derived from naturally-occurring ore deposits that were formed when seas and oceans evaporated. Mined potassium or potash is unearthed, purified and granulated with other nutrients. Typical potassium fertilizer sources are potassium nitrate, potassium sulfate, potassium chloride or mono-potassium phosphate.Significant potassium reserves exist around the world, but most production is done in Canada, Germany, Russia, Belarus and Israel.
- Secondary elements: Calcium (Ca) makes up 3.6 percent of the earth's crust, magnesium (Mg) makes up 2.1 percent and sulfur (S) is less than 1 percent. Calcium sources include limestone and gypsum rock, although lime and gypsum may be derived from industrial

processes. Fortunately for greenhouse and nursery growers, some irrigation waters can supply significant sources of calcium, magnesium and/or sulfate. Sphagnum peat moss can mineralize in a container and supply a significant supply of sulfur to ornamental crops.

- Micronutrients: Plants require boron (B), copper (Cu), iron (Fe), manganese (Mn), molybdenum (Mo) and zinc (Zn) in very small quantities on a parts per million (ppm) scale in plant tissue on a dry weight basis. These nutrients are derived from mineral ores whose presences in the earth's crust range from common to very rare. Copper, iron, manganese and zinc fertilizers are usually available in sulfated form or chelated to make them more available to plants over a wider range of root zone pH levels. Boron and molybdenum are available in a couple of forms of borates or molybdates.

Fertilizer Manufacturing

Fully integrated factories have been designed to produce compound fertilizers. Depending on the actual composition of the end product, the production process will differ from manufacturer to manufacturer.

Nitrogen Fertilizer Component

Ammonia is one nitrogen fertilizer component that can be synthesized from in-expensive raw materials. Since nitrogen makes up a significant portion of the earth's atmosphere, a process was developed to produce ammonia from air. In this process, natural gas and steam are pumped into a large vessel. Next, air is pumped into the system, and oxygen is removed by the burning of natural gas and steam. This leaves primarily nitrogen, hydrogen, and carbon dioxide. The carbon dioxide is removed and ammonia is produced by introducing an electric current into the system. Catalysts such as magnetite (Fe_3O_4) have been used to improve the speed and efficiency of ammonia synthesis. Any impurities are removed from the ammonia, and it is stored in tanks until it is further processed.

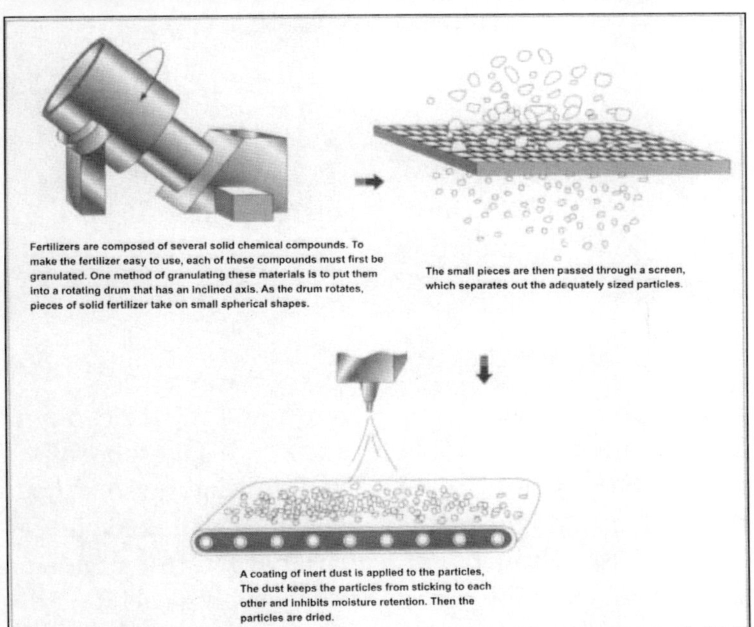

While ammonia itself is sometimes used as a fertilizer, it is often converted to other substances for ease of handling. Nitric acid is produced by first mixing ammonia and air in a tank. In the presence of a catalyst, a reaction occurs which converts the ammonia to nitric oxide. The nitric oxide is further reacted in the presence of water to produce nitric acid.

Nitric acid and ammonia are used to make ammonium nitrate. This material is a good fertilizer component because it has a high concentration of nitrogen. The two materials are mixed together in a tank and a neutralization reaction occurs, producing ammonium nitrate. This material can then be stored until it is ready to be granulated and blended with the other fertilizer components.

Phosphorous Fertilizer Component

To isolate phosphorus from phosphate rock, it is treated with sulfuric acid, producing phosphoric acid. Some of this material is reacted further with sulfuric acid and nitric acid to produce a triple superphosphate, an excellent source of phosphorous in solid form.

Some of the phosphoric acid is also reacted with ammonia in a separate tank. This reaction results in ammonium phosphate, another good primary fertilizer.

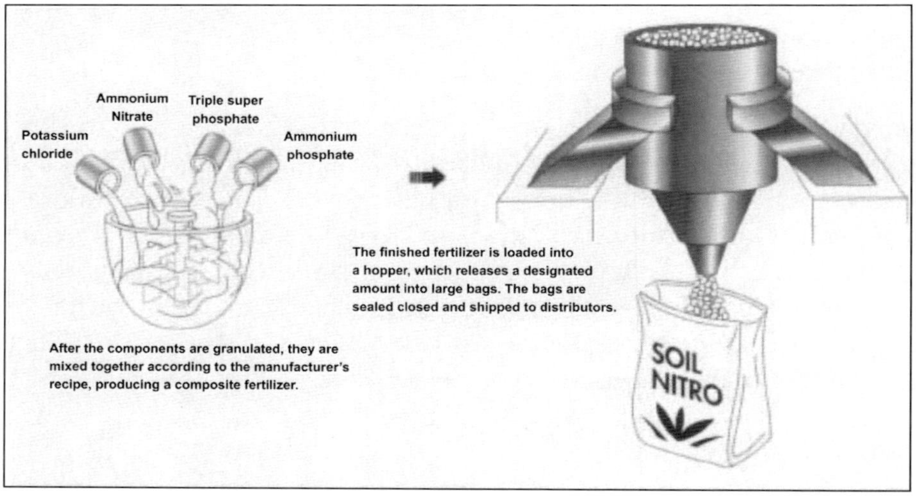

Potassium Fertilizer Component

Potassium chloride is typically supplied to fertilizer manufacturers in bulk. The manufacturer converts it into a more usable form by granulating it. This makes it easier to mix with other fertilizer components in the next step.

Granulating and Blending

To produce fertilizer in the most usable form, each of the different compounds, ammonium nitrate, potassium chloride, ammonium phosphate, and triple superphosphate are granulated and blended together. One method of granulation involves putting the solid materials into a rotating drum which has an inclined axis. As the drum rotates, pieces of the solid fertilizer take on small spherical shapes. They are passed through a screen that separates out adequately sized particles. A coating of inert dust is then applied to the particles, keeping each one discrete and inhibiting moisture retention. Finally, the particles are dried, completing the granulation process.

The different types of particles are blended together in appropriate proportions to produce a composite fertilizer. The blending is done in a large mixing drum that rotates a specific number of turns to produce the best mixture possible. After mixing, the fertilizer is emptied onto a conveyor belt, which transports it to the bagging machine.

Bagging

Fertilizers are typically supplied to farmers in large bags. To fill these bags the fertilizer is first delivered into a large hopper. An appropriate amount is released from the hopper into a bag that is held open by a clamping device. The bag is on a vibrating surface, which allows better packing. When filling is complete, the bag is transported upright to a machine that seals it closed. The bag is then conveyored to a palletizer, which stacks multiple bags, readying them for shipment to distributors and eventually to farmers.

Quality Contro

To ensure the quality of the fertilizer that is produced, manufacturers monitor the product at each stage of production. The raw materials and the finished products are all subjected to a battery of physical and chemical tests to show that they meet the specifications previously developed. Some of the characteristics that are tested include pH, appearance, density, and melting point. Since fertilizer production is governmentally regulated, composition analysis tests are run on samples to determine total nitrogen content, phosphate content, and other elements affecting the chemical composition. Various other tests are also performed, depending on the specific nature of the fertilizer composition.

Byproducts/Waste

A relatively small amount of the nitrogen contained in fertilizers applied to the soil is actually assimilated into the plants. Much is washed into surrounding bodies of water or filters into the groundwater. This has added significant amounts of nitrates to the water that is consumed by the public. Some medical studies have suggested that certain disorders of the urinary and kidney systems are a result of excessive nitrates in drinking water. It is also thought that this is particularly harmful for babies and could even be potentially carcinogenic.

The nitrates that are contained in fertilizers are not thought to be harmful in themselves. However, certain bacteria in the soil convert nitrates into nitrite ions. Research has shown that when nitrite ions are ingested, they can get into the bloodstream. There, they bond with hemoglobin, a protein that is responsible for storing oxygen. When a nitrite ion binds with hemoglobin, it loses its ability to store oxygen, resulting in serious health problems.

Nitrosamines are another potential byproduct of the nitrates in fertilizer. They are the result of a natural chemical reaction of nitrates. Nitrosamines have been shown to cause tumors in laboratory animals, feeding the fear that the same could happen in humans. There has, however, been no study that shows a link between fertilizer use and human tumors.

Cost of Creating Fertilizer

The majority of fertilizer raw materials are primarily used in field production agriculture worldwide.

The process of making these raw materials useful as fertilizers in greenhouse and nursery production costs money. These materials may have to be purified to remove components that could be injurious to ornamental container production or modified to fit production systems (e.g. be more soluble). They must then be transported to the blender plant, mixed together, packaged and shipped to the customer or distributor. Fertilizers are heavy, and as a result, transportation can be a major expense. Availability of shipping containers and trucks also has an impact on the ability to ship raw materials and finished goods in a timely fashion.

When you consider all the components that go into blended fertilizers, the origin of these materials, their limited worldwide supply and alternative uses, it's easy to see the inherent volatility of fertilizer pricing and availability. Although fertilizer is a fairly low percentage of a plant's total direct production costs, fertilizer prices will likely continue to rise with mounting oil prices and increasing demand for fertilizers in the global market.

Impact on Growers

Although fertilizers may only represent less than 5 percent of the total direct cost of producing a plant, a growing operation's annual fertilizer bill can add up to a significant sum. Since it's clear that fertilizer prices are unlikely to decrease any time soon, growers can do the following to reduce the potentially negative impact on their operations.

Switch to a lower-cost fertilizer: Upon first glance, switching to less expensive products may seem to be an easy choice. Yet, if these lower-grade fertilizers are made with poor-quality raw materials or offer inconsistent and unpredictable results, crop performance may suffer significantly.

Change formulations in order to use less fertilizer. Instead of only considering a fertilizer's cost per bag, it's better to compare the unit cost per crop when using different products. One strategy currently popular with growers is to choose products containing a higher percentage of nitrogen (N). Since water soluble fertilizers (WSF) are applied in concentration units of parts per million (ppm) of nitrogen, growers will need less product per gallon when they start with a formulation that contains a higher percentage of nitrogen.

However, other nutrients besides nitrogen are also important. It's possible to equalize nitrogen, but switching formulations might easily lead to a shortfall of another nutrient. Before making a change, study product labels carefully.

Use less of your current fertilizer. Some growers are successfully choosing to use the same fertilizer product while cutting their rates and concentrations dramatically. Current fertilizer programs may be providing more than adequate nutrition, so any extra is being wasted or leached out of the root zone.

However, reducing fertilizer rates may limit some nutrients such as micronutrients. One possible compromise is to reduce N-P-K, but supplement with micronutrients. This can be a cost-effective way to maintain crop quality.

Consider controlled-release fertilizers (CRF). Some industry professionals have suggested migrating away from CRFs to WSFs as a way to save money. However, a simple cost analysis demonstrates otherwise. While CRFs certainly are more expensive per pound compared to WSFs, they work more efficiently, can be applied at lower doses per pot and can result in less leaching of nutrients. Controlled-release fertilizers can easily be incorporated into the growing media before planting so the fertilizer is in close proximity to the roots where it is needed. Nutrient release from CRF is temperature-based, as is plant growth, so its use can be tied to plant demand. Plus, CRFs that are fully coated are not subject to huge nutrient losses caused by leaching. CRFs feed continuously as crops are watered, so root zones have a more uniform nutritional supply.

CRF rates for greenhouse use should be conservative. CRFs are much easier to adapt for crop types as compared to WSF, and their rates can be increased for heavy feeders without having to zone plants nutritionally. Longer-term CRFs can provide valuable post-production feeding both in the garden center and at the consumer's home, leading to better plant quality and customer satisfaction.

The best solution for dealing with fertilizer costs lies within the proper selection and use of the fertilizers you purchase. Fertilizer applications should provide the correct amounts and ratios of all essential nutrients at the right time to meet crop plant requirements. Any practice that doesn't achieve this basic goal will result in lower plant quality and, most likely, reduced revenues. The key is balancing fertilizer use with plant needs, which involves regular program assessment and an emphasis on best management practices. After all, you can use the best fertilizer in the world, but if it's not being used wisely or correctly, crop quality and your bottom line will suffer.

Application of Fertilizers

Broadcasting

Broadcasting is the method of application of fertilizer uniformly over the entire field. It may be at planting or in standing crop as top dressing.

This method is adopted under certain conditions:

a. Soils are highly deficient of nutrients, especially nitrogen.

b. Where fertilizers like basic slag, declaim phosphate, bone meal and rock phosphate etc. are to be applied to acid soils,

c. When potassic fertilizers are to be applied to potash deficient soils.

Top Dressing

Top dressing is application of fertilizer to the standing crop, especially nitrate nitrogenous fertilizers.

Placement

Inserting or drilling or placing the fertilizer below the soil surface by means of any tool or implement at desired depth to supply plant nutrients to crop either before sowing or in the standing crop is called placement. With placement methods, fertilizers are placed in the soil irrespective of the position of seed, seedling or growing plants before sowing or after sowing the crops.

The following methods are most common in this category:

1. Plough-sole Placement: The fertilizer is placed in a continuous band on the bottom of the furrow during the process of pouching. Each band is covered as the next furrow is turned. No attempt is usually made to sow the crop in any particular location with regard to the plough sole bands:

- This method has been recommended in areas where the soil becomes quite dry up to a few inches below the soil surface during the growing season, and especially with soils having a heavy clay pan a little below the plough-sole. By this method, fertilizer is placed in moist soil where it can become more available to growing plants during dry seasons.

2. Deep Placement: This method of application of nitrogenous and phosphates' fertilizers is adopted in paddy fields on a large scale in Japan and is also recommended in India.

- In this method, ammoniac nitrogenous fertilizer like ammonium sulphate or ammonium forming nitrogenous fertilizer like urea is placed in the reduced zone to avoid nitrogen loss in lowland rice and is available to the crop during the active vegetative period.
- Deep or sub-surface placement of the fertilizer also ensures better distribution in the root zone and prevents any loss by surface dram-off. Deep placement is done in different ways, depending upon the local cultivation practices.

In irrigated tracts, where the water supply is assured, the fertilizer is applied under the plough furrow in the dry soil before flooding the land and making it ready for transplanting. In areas where there is not too much of water in the field, it is broadcast before pudding. Paddling places the fertilizer deep into the root zone.

3. Sub-Soil Placement: This refers to the placement of fertilizers in the sub-soil with the help of heavy power machinery.

- This method is recommended in humid and sub-humid regions where many sub-soils are strongly acidic. Due to acidic conditions the level of available plant nutrients is extremely low. Under these conditions, fertilizers, especially phosphatic and potassic are placed in the sub-soil for better root development.

4. Localized Placement: This method refers to the application of fertilizers into the soil close to the seed or plant.

- Localized placement is usually employed when relatively small quantities of fertilizers are to be applied. Localized placement reduces fixation of phosphorus and potassium.

5. Bulk Blending: It is the process of mixing two or more different fertilizers varying in physical and chemical composition without any adverse effects.

- For this formulation certain additional materials called 'Fillers' and 'Conditioners' are used to improve the physical condition of the mixed fertilizer. This mixed fertilizer should be applied as top dressing.

6. Contact or Drill Placement: It refers to drilling seed and fertilizer simultaneously at sowing.

- Ferti seed drills are popular in dry land agriculture for drill placement.

7. Band Placement: It consists of applying the fertilizer in continuous or discontinuous bands close to the one or both sides of seed or plant.

- This method is well suited for wider spaced crops i.e. sugarcane, tobacco, cotton, castor, maize and vegetables.

8. Pellet Placement: Application of fertilizer, especially nitrogen in pellet form to the lowland rice to avoid nitrogen loss.

- Soil and fertilizer are mixed in the ratio of 1; 10 or 15 in the form of small pellets and placed in the reduced zone.

Liquid Fertilization

The use of liquid fertilizers as a means of fertilization has assumed considerable importance in foreign countries.

Starter Solution

Solutions of fertilizers, generally prepared in low concentration.

- It consists of N_2,P_2O_5, and K_2O in the ratio of 1:2:1 and 1:1:2 are applied for soaking seed, dipping roots or spraying on seedlings for early establishment and growth.
- The nutrients reach the plant roots immediately.
- The solution is sufficiently diluted so that it does not inhibit growth.
- This method is specially applied in pulses and vegetable crops.

Foliar Spray

In this method, nutrients are applied to the standing crops in the form of spray for quick recovery from the deficiency. It avoids fixation of nutrients in the soil.

Foliar application does not result in a great saving of fertilizer but it may be preferred under the following conditions:

- When visual symptoms of nutrient deficiencies observed during early stages of deficiency.
- When unfavorable soil physical and chemical conditions, which reduce fertilizer use efficiency (FUE).
- During drought period where in the soil application could not be done for want of soil moisture.

Soil Application

Direct application of liquid fertilizers to the soil need special injecting equipment. Anhydrous ammonia (a liquid under high pressure up to 14 kg per square cm. or more) and nitrogen solutions are directly applied to the soil.

- Plant injury or wastage of ammonia is very little if the material is applied about 10 cm below the seed. If the application is shallow, nitrogen from ammonia will be lost. This method allows direct utilization of the cheapest nitrogen source.
- Liquid manures such as urine, sewage water and shed washing are directly let into the field.

Fertigation

Fertigation is the application of fertilizer with irrigation water in either open or closed system. The open system includes lined and unlined open ditches and gated pipes that are used for furrow and flood irrigation. Sprinkler and drip systems are the main closed systems.

- Nitrogen and sculpture are the principle nutrients applied by Fertigation.
- This saves the application cost and allows the utilization of relatively in expensive water-soluble fertilizers.

General Absorption and Mobility Rankings for foliar Applied Nutrients.	
Absorption	Mobility
Rapid	Mobile
Urea Nitrogen, Potassium, Zinc	Urea Nitrogen, Potassium, Phosphorus, sulfate
Moderate	Partially Mobile
Calcium, Sulfate, Manganese Boron	Zinc, Copper. Manganese Boron Molybdenum
Slow	Immobile
Magnesium, Copper, Iron,	Iron Calcium Magnesium
Molybdenum	
Nitrification Inhibitors For Lowland: Oxamide (31 percent N) & thiorea (36.8percent N) For upland: N-serve Others: Nitropyrin, Neem cake (Mimbiolin)	

Slow Released N Fertilizers

- Urea formaldehyde (UF – Urea form): 38-42 per cent N.
- Ox-amide.
- Sulpher coated urea, Neem coated urea.
- Formation of super granule (Modified form): Urea 1-4 gm. of big granules.

References

- Fertilizer: britannica.com, Retrieved 16 July, 2019
- Agricultural-technology, technology: britannica.com, Retrieved 18 August, 2019
- Fertilizers: chemicalsafetyfacts.org, Retrieved 20 January, 2019

- Fertilizers, biology: byjus.com, Retrieved 23 Feb February, 2019
- Fertilizers, materials-and-applications: essentialchemicalindustry.org, Retrieved 24 March, 2019
- Nitrogen: cropnutrition.com, Retrieved 26 April, 2019
- Phosphate-fertilizers, fertilizers, commodities: aithercommodities.com, Retrieved 28 May, 2019
- Understanding-phosphorus-fertilizers, chemical-composition, phosphorus-and-potassium: umn.edu Retrieved 30 June, 2019
- Phosphate-fertilizers-manufacturing-process-of-phosphate-fertilizers: guichon-valves.com, Retrieved 1 July, 2019
- Potassium: cropnutrition.com, Retrieved 3 August, 2019
- What-is-potash: cropnutrition.com, Retrieved 8 January, 2019
- Facts-about-fertilizer-raw-materials-availability-and-pricing: gpnmag.com, Retrieved 3 February, 2019
- Fertilizer: madehow.com, Retrieved 5 March, 2019
- Facts-about-fertilizer-raw-materials-availability-and-pricing: gpnmag.com, Retrieved 7 April, 2019
- Methods-of-application-of-fertilizers-soil-science, fertilizers: yourarticlelibrary.com, Retrieved 9 May, 2019

Chapter 5

Nutrient Management and Soil Health

The science and practice of achieving optimal nutrient use efficiency, crop quality, economic returns, etc. while reducing off-site transport of nutrients is known as nutrient management. Soil health is a condition where a soil meets its range of ecosystem functions as appropriate to its environment. The chapter closely examines the key concepts of nutrition management and soil health to provide an extensive understanding of the subject.

Nutrient management is the process of managing the amount, source, timing, and method of nutrient application with the goal of optimizing farm productivity while minimizing nutrient losses that could create nutrient (biosolid) application environmental problems. It includes developing nutrient budgets that consist of knowing the amounts of nutrients present in the soil, determining the amount of nutrients needed by the crop, accounting for all the potential sources of nutrients, and then applying manures, composts, irrigation water, or inorganic fertilizers to meet the nutrient need of the crop. It also uses site management practices to increase or maintain soil quality to reduce the potential for erosion and nutrient transport into surface water or nutrient leaching into groundwater. Soil quality is an important component of nutrient management because it affects nutrient retention and water movement through the soil. Biosolids are applied to a crop field in the photo at right.

Importance of using Nutrient Management

Farmers need to apply nitrogen, phosphorus, potassium and other nutrients to achieve desired crop growth and yield. However, excessive nutrient application can have negative environmental impacts. Nutrients that are not efficiently used by crops or retained in the soil can leach into groundwater and move from agricultural land into surface waters. For example, excess nitrogen in the form of nitrate can leach through the soil into groundwater. Concentrations of nitrate in drinking water above the national drinking water standard of 10 mg/L can cause health problems for infants and elderly people. Nitrate-nitrogen concentrations of 20 to 40 ppm or mg/L can cause health problems for horses and ruminants.

Excess nitrogen and phosphorus applied to crops can move into surface water with runoff during storm events causing nutrient enrichment (eutrophication) of these water bodies, as shown in the photo at right. Eutrophication causes excessive plant and algae growth that causes the water to turn green. When the plants and algae die, the organisms that decompose the dead material consume much of the oxygen in the water. This can cause fish kills or shifts in the types of fish species present, and can prevent recreational uses of the water. Although excess phosphorus is usually the nutrient that causes eutrophication issues in fresh water reservoirs, excess nitrogen (concentrations greater than 1 mg/L) is generally the nutrient that causes eutrophication problems in coastal waters.

Although many people do not think about the relationship between soil quality and water quality, the link between them is strong. Good soil quality is critical to protecting water quality by functioning to hold water, adsorb nutrients, and retain other contaminants. For a soil to perform these functions, its capacity to absorb nutrients cannot be exceeded. Nutrient management is critical to maintain adequate, but not excessive nutrient concentrations for crop production and maintaining soil quality.

A schematic of the phosphorus (P) cycle Illustrating how water running off the soil surface can move phosphorus into nearby streams or ponds and how good soil management can reduce those runoff losses

Note the higher amount of soil organic matter in the schematic on the left hand side of the stream and the lower particulate and soluble P losses. The size of the arrows represents the magnitude of P input, recycled or lost from the system. From Nutrient Cycles in the Southern Piedmont. 1998. Dr. Dory Franklin (ed). USDA ARS, Watkinsville, GA.

In addition to the environmental benefits, there are economic benefits to nutrient management. Evaluating what nutrients you need for an expected crop production yield and accounting for nutrients provided by the soil, manures, composts, or legumes can reduce the supplementary amount of inorganic fertilizer needed.

Many farmers use a nutrient management plan to help them make nutrient input and economic decisions. In some areas and for some crops, the nitrogen need is based on a "realistic yield goal". This yield goal should be based on yield averages from your farm for at least three to five years. If farm yields are not available, county averages can be used.

A large part of nutrient management is record keeping. Record keeping, along with calibration of application equipment can insure proper application rates. Keeping records of all nutrients applied, cost of nutrients, crop yields, and livestock productivity will help determine what is profitable but more importantly what is not profitable for your operation. This process also helps identify what modifications should be made to improve productivity.

Why doesn't Everybody use Nutrient Management?

Most farmers use some form of nutrient management. Using soil tests to determine the amount of inorganic fertilizer needed for a crop is a basic step of nutrient management. As the price of inorganic fertilizer increases, more people are paying close attention to nutrient sources such as legumes, manures, composts, or other byproducts to help reduce their input costs. Many producers find the annual variations in yield, crop response, climate, soil types, and manager decision confusing, but well kept records and field experience help understand and explain these variations.

How is Soil Affected?

The key to good soil quality is soil organic matter. A sufficient amount of nutrients in the soil, particularly nitrogen, is necessary to form and maintain soil organic matter. A fertile soil has greater plant growth, which can create greater inputs of roots and other plant debris into the soil. This plant debris undergoes decomposition and adds to the soil organic matter. Applications of animal manures and composts, as well as the use of cover crops, all help increase soil organic matter. Organic matter provides a food source for soil microbes and increases microbial activity. As the microbes breakdown organic matter, nutrients are released in forms that the plants can utilize. Because nutrient management accounts for the nutrients added to the system, it promotes increasing soil quality without creating nutrient excesses.

Different soils have different capacities to adsorb and retain nutrients. This is related to the amount of soil organic matter and the soil texture (percent sand, silt, clay). Because the soil texture cannot be changed, increasing soil organic matter is the best way to increase the capacity of a soil to retain nutrients. Soils with larger amounts of soil organic matter and at a near neutral pH will have a greater capacity to retain nutrients, thus a higher soil pH affect on nutrient availabilityquality than soils with low organic matter.

Soil pH also affects the availability of nutrients in the soil. A soil pH around 6.5 allows for maximum availability of the soil nutrients and microbial activity. Poultry and dairy manure have high levels of calcium and a natural liming affect on the soil, which can provide an additional benefit to acidic soils. Soil testing for nutrient management helps farmers track changes in pH so they know when additional lime is needed and can maintain optimal pH and nutrient availability.

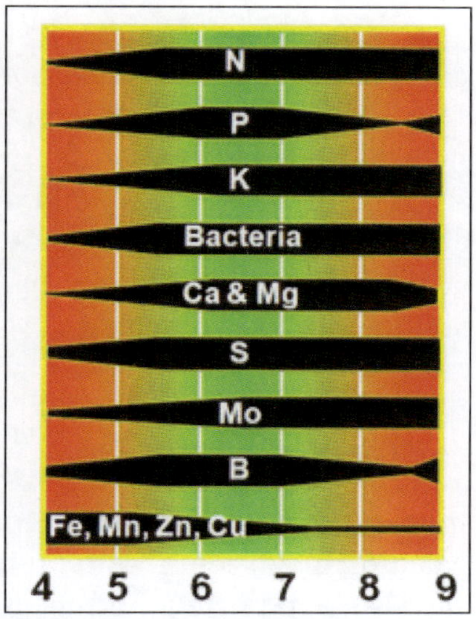

Effect of soil pH on nutrient availability. Nutrients reach maximum availability near pH 6.5.

Although soil organic matter can also retain some negatively charged nutrients such as phosphorus, the clay and carbonate content of the soil has a greater impact on nutrient retention. Aluminum and iron oxides found in clayey soils readily bind phosphorus; consequently, soils with greater percent clay can retain higher concentrations of phosphorus than other soil types. Soils with high carbonates also tend to bind phosphorus and reduce its loss. Sandy soils have fewer iron and aluminum oxides, therefore, have a lower capacity to retain phosphorus. If phosphorus is over-applied (which can happen when manures are used to meet the nitrogen need of a crop), the capacity of the soil to bind phosphorus can be exceeded. In these cases, the soil no longer functions as a nutrient buffer, soil quality is reduced, and water quality can be impacted.

Because soil quality affects how water moves into surface and groundwater, it plays an important role in the site management practices of a nutrient management plan. Soil organic matter helps sustain microbial activity, which in turn tends to create soil aggregates by generating polysaccharides and other compounds that "glue" soil particles together. These aggregates can help prevent soil crusting and promote better soil structure, which leads to easier root penetration, as well as improved plant growth and production. Within a soil textural class, higher soil organic matter and better aggregation will allow more water to infiltrate into the soil, reducing erosion and preventing the loss of nutrients.

In areas where wind erosion is a problem, the aggregation of soil particles by soil organic matter, can also help reduce soil loss due to wind. Excessive tillage and the lack of residues or cover crops decrease soil organic matter and increase both wind and water erosion. Part of a nutrient management plan might be to change management practices to build or maintain soil organic matter to reduce erosion or establishing buffers to remove nutrients as water moves through the buffer. The management changes will be specific to the soil type, slope, and crops grown. The photo at right shows a conservation buffer adjacent to a crop field in Iowa.

How to Monitor the Soil

The best way to determine the nutrient content of the soil is by testing. Soils should be tested every 1-3 years based on the soil type and state requirements or recommendations. Soil tests usually report pH, phosphorus, potassium, calcium, magnesium, sulfur, and micronutrients. In some states, the level of soil nitrate nitrogen may also be reported. From soil analyses, nutrient recommendations can be offered based on local field trials and the experience of the land grant university. All this information helps the farmer know how much lime and fertilizer is needed for a particular crop on a particular soil.

Although soil organic matter is an important indicator for soil quality, it is often not reported in a regular soil test report. Most laboratories can analyze soil organic matter or soil carbon. Asking for this analysis is helpful to determine if soil organic matter is increasing, decreasing or staying the same soil.

For the best soil test results, the in-field soil sampling must be conducted properly. Individual fields should be sampled separately; areas within a field that are managed differently should also be sampled separately. In fields with different soil types each soil should be sampled separately. Cores should be taken in a zig-zag pattern across the sample area. Sampling in this way ensures a representative sample across the sample area will be collected. Soil samples should be taken from 0 to 6 inches for row crops and 0 to 4 inches in pastures and hayfields. A minimum of 15 cores should be taken from the areas that are less than 15 acres. For sample areas greater than 15 acres, a good rule of thumb is one soil core per acre. Refer to your local land grant university for appropriate soil sampling protocol. The cores taken from each sampling area should be mixed thoroughly in a plastic bucket and combined into one representative sample as shown in the photo at right. A minimum of 1 pint of soil should be collected for each sampling area. Check with your local NRCS or Cooperative Extension for details on recommended practices for soil sampling in your area.

Local Considerations

The specifics of a nutrient management plan are dependent on soil types, cropping systems, and site conditions such as topography and hydrology. Each state has their own specific procedures for developing a nutrient management plan based on their research and experience. It is recommended that you contact your local cooperative extension or NRCS office to receive specific guidance for your area.

Managing nutrients with inorganic fertilizers is relatively easy because nutrients can be specifically blended in the concentrations needed for a particular crop. Nutrient management can be more difficult when organic fertilizers such as manures or composts are used. This is because the nitrogen to phosphorus (N:P) concentrations in organic fertilizers tend to be around 1:1 which does not match the N:P need of most crops which is 4-6:1. Consequently, when manures are regularly applied to meet the nitrogen needs of the crop, phosphorus is over-applied.

Composts can create similar conditions, especially when manures are used as a feedstock. Composting causes nitrogen concentrations in the organic material to decrease because some of the original nitrogen is lost as ammonia gas. Phosphorus is concentrated because the volume of the material decreases during composting and it does not have a gaseous from. Consequently, the N:P ratio in compost does not match plant requirements and phosphorus can be over-applied with regular use.

Elements of a Nutrient Management Plan

Nutrient management consists of several steps:

- Testing the soil to determine the nutrient supplying power of the soil,
- Determining the recommended amounts of nutrients needed to produce the desired yields,
- Accounting for nutrient inputs from other sources, such as legumes,
- Analyzing manures, composts, and irrigation water to determine the nutrient content,

- Applying manures or composts at recommended rates and based on the critical nutrient (usually either nitrogen or phosphorus),
- Applying the additional inorganic nutrients as needed,
- Keeping records so evaluations and adjustments can be made, and
- Being aware of your surrounding landscape so sensitive areas can be protected.

This process is an important step in building and maintaining soil quality.

Paddy requires the following essential nutrients for its normal development.

Carbon	Nitrogen	Calcium
Hydrogen	Phosphorus	Magnesium
Oxygen	Potassium	Sulphur
Iron	Zinc	Chlorine
Manganese	Boron	
Copper	Molybdenum	

Nitrogen, phosphorus and potassium are known as primary plant nutrients; calcium, magnesium and sulphur, as secondary nutrients; iron manganese, copper, zinc, boron, molybdenum and chlorine as trace elements or micro-nutrients. The primary and secondary nutrient elements are known as major elements. This classification is based on their relative abundance, and not on their relative importance. The micronutrients are required in small quantities, but they are important as the major elements in plant nutrition.

Nitrogen

- Nitrogen, the most important nutrient for rice, is universally limiting the rice productivity.
- Nitrogen encourages the vegetative development of plants by imparting a healthy green color to the leaves.
- It seems that majority of Indica varieties are adapted to relatively low levels of nitrogen in the region of 25 kg N/ha.
- Rice plant depends mainly for its nitrogen upon the decomposition of organic matter under anaerobic conditions and in the early stages of growth takes up nitrogen in the form of ammonia which is the stable form of nitrogen in submerged soils.
- There are two stages in the growth of rice crop when nitrogen is most needed; early vegetative and panicle initiation stages.
- Fertilizing the crop during early vegetative growth promotes tillering leading to higher yield. Application at panicle initiation or early booting stage will help the plant produce more and heavier grains per panicle.

Encourages the vegetative growth of paddy

Phosphorus

1. Phosphorus is particularly important in early growth stages.
2. It is mobile within the plant and promotes root development (Particularly the development of fibrous roots),tillering and early flowering.
3. Addition of mineral P fertilizer is required when the rice plants root system is not yet fully developed and the native soil P supply is inadequate.
4. Phosphorus is remobilized within the plant during later growth stages if sufficient P has been absorbed during early growth.
5. It also increases resistance to disease and strengthens the stems of cereal plants, thus reducing their tendency to lodge. It offsets the harmful effects of excess nitrogen in the plant.

Promotes fibrous root development

Potassium

1. Potassium enhances the ability of the plants to resist diseases, insect attacks, cold and other adverse conditions.
2. It plays an essential part in the formation of starch and in the production and translocation of sugars, and is thus of special value to carbohydrate-rich crops.
3. Involves in working of enzymes.

4. Helps in production and movement of photosynthates to sink.
5. Helps in proper uptake of other nutrients.
6. Influences tillering or branching of plant and size and weight of grain.
7. Over 80 per cent of the absorbed potassium by the plant is found in straw. Need for potassium is most likely to occur on sandy soils.

Gives resistance in adverse conditions

Calcium

1. Calcium combines with pectin in the plant to form calcium pectate, which is an essential constituent of the cell-wall.
2. It also promotes the activity of soil bacteria concerned with the fixation of free nitrogen or the formation of nitrates from organic forms of nitrogen.
3. Furthermore, it is necessary for the development of a good root system.

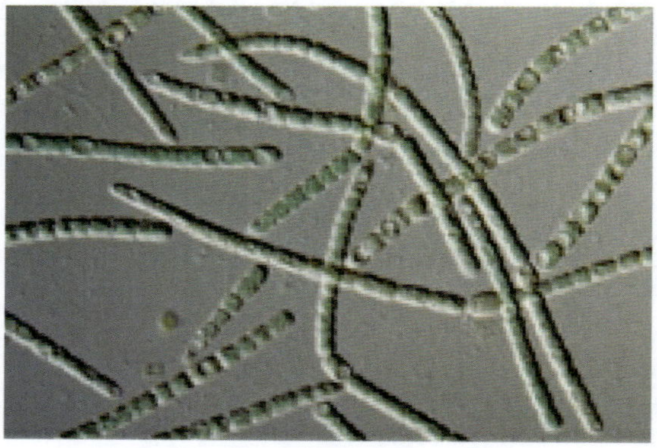

Promotes the activity of soil bacteria

Magnesium

1. Magnesium is an essential constituent of chlorophyll.
2. It is usually needed by plant in relatively small quantities. Hence its deficiency in the soil is experienced later than that of potassium.

Essential constituent of chlorophyll

Sulphur

1. It involve in chlorophyll production, protein synthesis and plant function and structure.
2. Sulphur forms an important constituent of straw and plant stalks.

Constituent of straw and stalk

Iron

1. Iron is necessary for the synthesis of chlorophyll.
2. Mainly a problem in upland soils.

Necessary for chlorophyll synthesis

Zinc

1. Essential for the transformation of carbohydrates.
2. Regulates consumption of sugars.
3. The function of zinc in plants is as a metal activator of enzymes.
4. Deficiency of zinc in lowland rice occurs in near neural to alkaline soils, particularly in calcareous soils.
5. Availability of both soil and applied zinc is higher in upland soil than in submerged soil.
6. Soil submergence causes decrease in zinc concentration in the soil solution.
7. Rice crop removes 30-40 g Zn per tonne of grain.

Helps in uptake of nutrients

Boron

1. Boron facilitates the translocation of sugars by forming sugar borate complex.
2. It involves in cell differentiation and development since boron is essential for DNA synthesis.
3. Also involves in fertilization, hormone metabolism etc.

Boron helps in fertilization

Copper

1. It is an important constituent of plastocyanin (copper containing protein).
2. It is also a constituent of several oxidizing enzymes.
3. Important for reproductive growth.
4. Aids in root metabolism and helps in the utilization of proteins.

Important for panicle development

Manganese

1. It is an activator of nitrite reductase and many respiratory enzymes.
2. It is necessary for the evolution of oxygen (photolysis) during photosynthesis.
3. Functions with enzyme systems involved in breakdown of carbohydrates, and nitrogen metabolism.
4. Soil is a source of manganese.

Helps in uptake of nitrogen

Silicon

1. It is an important element for improving plant health and disease resistance.
2. It has the potential to significantly decrease the susceptibility of certain plants to both biotic and abiotic diseases.
3. Beneficial for the growth of paddy.

4. An adequate supply of silica is essential for paddy to give a good yield by increasing the strength and rigidity of cells.
5. Plant roots take up their silica as silicic acid.
6. It is possible that the amount taken up by paddy is equal to the amount of silicic acid present in the water the roots absorb, so that the greater the amount of water transpired, the greater their uptake of silica.

Gives disease resistance

Nutritional Disorders

Nitrogen

Nitrogen deficiency

Delayed Maturity

LCC

Deficiency Symptoms

1. Stunted growth and yellowish plants.
2. Older leaves of whole plants are yellowish green.
3. Old leaves sometimes all leaves become light green.
4. Deficiency symptoms first appear at the tip and progress along the midrib until the entire leaf is dead.
5. Tips become chlorotic. Leaves narrow, short, erect and lemon-yellowish green.

Corrective Measures

1. Do not apply large amounts of N to less responsive varieties.
2. Choose suitable plant spacing for each cultivar.
3. Adjust the number of splits and timing of N applications according to the crop establishment method.
4. Keep the field flooded to prevent denitrification but avoid N losses from water runoff over bunds immediately after fertilizer application.
5. Soil application of 25% excess of recommended N.
6. Foliar application of Urea 1% at weekly interval till the symptoms disappear.

N Management in Paddy with Leaf Colour Chart (LCC)

1. Yellowing of leaves in paddy indicates the deficiency of nitrogen, but it is difficult to decide the quantity of nitrogen to be applied based on the extent of yellowing.
2. The actual requirement of nitrogen by the crop can be correctly assessed by leaf colour chart.
3. The leaf colour chart consists of 6 -7 green strips, 1st strip with light green colour and the last strip (6th or 7th) with dark green colour, and in between strips (2nd to 5th) are with varying intensity of green colour.

Use of Leaf Colour Chart

1. Select fully opened disease free new leaf i.e. third leaf from the top as index leaf in paddy plant for assessing the leaf colour and ten leaves to be selected from ten plants in the field.
2. Match the colour of the selected leaves by keeping the middle of the leaves on the colour strips of leaf colour chart and assess the colour intensity (LCC value) during morning hours (8-10 am).
3. Assess the intensity of leaf colour each time at a particular time by a particular individual.
4. Take average of two if the leaf colour matches between two colour strips of the chart.

5. Commence the assessment of the leaf colour with LCC at 14 DAT in transplanted rice or 21 DAS in direct seeded rice and continue up to flower initiation/heading at an interval of 7-10 days.

6. Critical LCC value varies with the type of paddy genotypes. LCC critical value is 3.0 in low N response cultures. In Tamilnadu LCC critical value is 3.0 in low N response cultures like White Ponni and 4.0 in other cultivars and hybrids.

7. Assess the average LCC values of 10 leaf samples. When the average LCC value of ten leaves or when the LCC values of five or more leaves found below the critical LCC limit fixed for that genotype, then top dress nitrogen depending on the crop growth and stage.

8. If six or more leaves read below the specified threshold value, N can be applied 35 kg N/ha in dry season and 30 kg N/ha in wet season per application per ha. If the value is above the threshold value, there is no need for top dressing during that week.

Nitrogen Toxicity Symptoms

1. Plants are dark green in colour.
2. Abundant foliage.
3. Restricted root system.
4. Flowering and seed setting may be retarded.

Nitrogen Sources

Farmyard manure, Green manures, Biofertilizers (Rhizobium, AzollaAzospirillum, Azotobacter) Castor cake, Neem cake, Urea, Diammonium phosphate, Ammonium nitrate, Ammonium sulphate.

Phosphorus

Deficiency Symptoms

1. Plants stunted with reduced tillering.
2. Leaves narrow, short, very erect 'dirty' dark green.
3. Older leaves turn brownish red and purple colors develop in leaves.
4. Stems thin and spindly.
5. Poor tillering/branching.
6. Poor root growth.

Corrective Measures

1. Application of phosphobacteria to the soil as seed coating or as seedling dip.

2. Application on P fertilizer 15-30 kg P/ha.
3. Rock phosphate broadcast before flooding when soil pH is low.

Phosphorus Sources

Farmyard manure, Biofertilizers (Phosphate solubilizers) Castor cake, Neem cake,Super phosphate (single), Super phosphate (Double), Super phosphate (Triple), Basic slag, Mussori, Diammonium phosphate (SPIC), Ammonium phosphate (Gromor).

Phosphorus Toxicity (Injury due to Excess P Application)

1. Fixed in soil – not available to plants.
2. Leads to Zinc deficiency.

Phosphorus deficiency

Phosphorus toxicity

Potassium

Deficiency Symptoms

1. Dark green plants with yellowish.
2. Brown leaf margins and brown necrotic spots on the tips of older leaves.
3. Rusty brown spots on the panicles and poor grain formation.
4. Weak stem leads to lodging.

Corrective Measures

1. Soil application of 25% excess of recommended K.
2. Foliar application of 1% KCL.
3. Spread and incorporate the straw evenly over the field before burning. Ash from burnt straw heaps should also be spread over the field.

Potassium Sources

Farmyard manure, Castor cake, Neem cake, Muriate of potash (KCl), Potassium sulphate.

Injury due to Excess K Application

1. Excess of this element tends to delay maturity.
2. Leads to calcium magnesium and iron deficiencies.

Potassium deficiency

Potassium toxicity

Calcium

Deficiency Symptoms

1. White or bleached, rolled, and curled tips of youngest leaves.
2. Necrosis along the lateral margins of leaves.
3. Old leaves turn brown and die.
4. Stunting and death of growing points.

Corrective Measures

Calcium deficiency

1. Apply farmyard manure or straw (incorporated or burned) to balance Ca removal in soils containing small concentrations of Calcium Fertilizers.
2. Apply $CaCl_2$ or Ca containing foliar sprays for rapid treatment of severe Ca deficiency.
3. Apply gypsum in Ca deficient high pH soils, e.g., on sodic & high K soils.
4. Apply lime on acid soils to raise pH and Ca availability.
5. Apply pyrites to mitigate the effects of $NaHCO_3^-$ rich water on Ca uptake.

Calcium Sources

Farmyard manure, Calcium chloride. Gypsum, dolomite, lime, pyrites, single superphosphate or triple superphosphate.

Magnesium

Deficiency Symptoms

1. Leaf chlorotic with white tips.
2. Pale-colored plants with orange-yellow interveinal chlorosis on older leaves and later on younger leaves.
3. Chlorosis progresses to yellowing and finally necrosis in older leaves in severe cases wavy and droopy leaves.
4. Reduced number of spikelets and grain quality.

Corrective Measures

Rapid correction of Mg deficiency symptoms is achieved by applying a soluble Mg source such as kieserite or Mg chloride.

Foliar application of liquid fertilizers containing Mg (e.g.,Mgcl 22%).

Magnesium Sources

Farmyard manure, Magnesium chloride, dolomite.

Magnesium deficiency

Sulphur

Deficiency Symptoms

1. Yellowing or pale green whole plant.
2. Young leaves chlorotic or light green colored with the tips becoming necrotic.
3. Lower leaves not showing necrosis.
4. Leaves pale yellow.
5. Effect on yield is more pronounced when S deficiency occurs during vegetative growth.

Corrective Measures

1. Incorporate straw instead of completely removing or burning it. About 40-60% of the S contained in straw is lost during burning.
2. Carry out dry tillage after harvesting, to increase the rate of sulfide oxidation during the follow period.
3. Applying 15-20 kg S ha-1 gives a residual effect that can supply the S needed for two subsequent rice crops.

Sulphur Sources

Ammonium sulfate, Single superphosphate, Potassium sulfate, gypsum and S-coated urea.

Sulfar deficiency

Sulfide toxicity-chlorosis of young leaves

Iron

Deficiency Symptoms

1. Interveinal yellowing.
2. Chlorosis of whole leaves and emerging leaves.
3. Entire plants becomes chlorotic.

Corrective Measures

1. Apply solid $FeSO_4$ (30 kg Fe/ha) next to rice rows or broadcast.
2. Foliar applications of $FeSO_4$ (2-3% solution) 2-3 applications at 2 week intervals.
3. Use acidifying fertilizers (e.g., ammonium sulfate instead of urea) on high-pH soils.
4. Grow tolerant cultivars for low soil Fe availability.

Iron Sources

Soluble ferrous sulfate (20-33% Fe), ferrous ammonium sulfate (14% Fe), and iron chelates (5 to 14% Fe).

Iron deficiency

Iron toxicity

Zinc

Deficiency Symptoms

1. Dusty brown spots on upper leaves.
2. Stunted growth of plants.
3. Decreases tillering and increase spikelet sterility.
4. Leaf base of younger leaves become chlorotic brown and blotches/streaks on lower leaves.

Corrective Measures

1. Broadcast $ZnSO_4$ in nursery seedbed.
2. Dip seedlings or presoak seeds in 2-4% ZnO suspension.
3. It is enough to apply 12.5 kg zinc sulphate /ha, if green manure (6.25 t/ha) or enriched FYM, is applied.
4. Apply 25 kg of zinc sulphate with 50 kg sand before transplanting.
5. Apply 5-10 kg Zn ha-1 as Zn sulfate, apply 0.5 – 1.5 % $ZnSO_4$/ha as a foliar spray at tillering (25-30 DAT), 2-3 repeated applications at intervals of 10-14 days.
6. Zn chelates (e.g., Zn-EDTA) can be used for foliar application.

Zinc deficiency

Zinc toxicity

Zinc Sources

Zinc sulfate, Zinc carbonate, Zinc chloride, Zinc chelate, Zinc oxide.

Zinc Toxicity Symptoms

Excess zinc commonly produces iron chlorosis in plants.

Aluminium

Toxicity Symptoms

1. Orange-yellow interveinal chlorosis on leaves followed by leaf tip death and leaf margin scorch.
2. Necrosis of chlorotic areas during severe Al toxicity.
3. Stunted and deformed roots in susceptible cultivars.

Corrective Measures

1. Delay planting until pH has increased sufficiently after flooding (to immobilize Al).
2. Plant Al-tolerant cultivars (IR43, CO 37, and Basmati 370) which accumulate less Al in their foliage and take up and use Ca and P efficiently in the presence of Al.
3. Apply 1-3 t lime/ha to raise pH.

Boron

Deficiency Symptoms

1. White and rolled leaf tips of young leaves.
2. Reduction in plant height.
3. Death of growing points, but new tillers continue to emerge during severe deficiency.
4. Plants unable to produce panicles if affected by B deficiency.

Corrective Measure

1. Avoid excessive leaching (percolation).
2. Apply B in soluble forms (borax) for rapid treatment of B deficiency (0.5-3 kg B/ ha), broadcast and incorporated before planting, top dressed or as foliar spray during vegetative rice growth.

Boron Sources

Anhydrous borax, Fertilizer borate, borax.

Toxicity Symptoms

1. Chlorosis of tips and margins of older leaves as initial symptoms.
2. Dark brown elliptical spots on discolored areas two to three weeks later followed by browning and drying up.
3. Necrotic spots prominent at panicle initiation.

Corrective Measures

1. Plant B-toxicity tolerant varieties (e.g., IR42, IR46, IR48, and IR 54).
2. Use surface water with a low B content for irrigation.
3. Plough when the soil is dry so that B accumulates in the topsoil. Leach with water containing a small amount of B.

Boron deficiency

Boron toxicity

Copper

Deficiency Symptoms

1. Leaves develop chlorotic streaks on either side of the midrib.
2. Dark brown necrotic lesions on leaf tips, Leaves often bluish green and chlorotic near the leaf tip.
3. New leaves do not unroll and the distal parts of leaves maintain a needle like appearance.
4. Reduced tillering and increased spikelet sterility.

Corrective Measures

1. Dip seedling roots in 1 $CuSO_4$ suspensions for 1 hr before transplanting.
2. Avoid overliming of acid soils because it may reduce Cu uptake.
3. On Cu-deficient soils, apply CuO or $CuSO_4$ (5-10 kg Cu/ha at 5-year intervals) for long-term maintenance of available soil Cu (broadcast and incorporate in soil).

Copper Sources

Cupric sulfate, Cu oxide.

Copper deficiency

Manganese

Deficiency Symptoms

1. Pale greyish green interveinal chlorosis spreads from the tip to the leaf base.
2. Necrotic brown spots develop later and leaf becomes dark brown.
3. Newly emerging leaves short, narrow and light green.
4. Deficient plants shorter, with fewer leaves, weigh less, and smaller root system at tillering.

Corrective Measures

1. Apply $MnSO_4$ or finely ground MnO (5-20 kg Mn/ha) in bunds along rice rows.
2. Apply foliar $MnSO_4$ for rapid treatment of Mn deficiency(1-5 kg Mn/ha in about 200 L water/ha).
3. Use acid-forming fertilizers, e.g., ammonia sulfate [$(NH_4)_2SO_4$] instead of urea.

Manganese Sources: Mn sulfate, Mn chloride

Toxicity Symptoms

1. Yellowish brown spots between leaf veins, extending to the whole interveinal area.
2. Brown spots on veins of lower leaf blades and leaf sheaths.

3. Leaf tips dry out eight weeks after planting.
4. Chlorosis of younger (upper) leaves, with symptoms similar to those of Fe chlorosis.
5. Stunted plants, reduced tillering and sterility results in reduced grain yield.

Corrective Measures

1. Coat seeds with oxidants (e.g., Ca peroxide) to improve germination and seedling emergence by increasing the supply of O_2.
2. Apply lime on acid soils to reduce the concentration of active Mn.
3. Mn uptake is reduced in the presence of ammonia.

Manganese deficiency

Manganese toxicity

Silicon

Deficiency Symptoms

1. Leaves and culms become soft and droopy thus increasing mutual shading.
2. Reduces photosynthetic activity.
3. Severe Si deficiency reduces the number of panicles m2 and the number of filled spikelets per panicle.
4. Si-deficient plants are particularly susceptible to lodging.

Corrective Measures

1. For more rapid correction of Si deficiency, granular silicate fertilizers should be applied: Calcium silicate 120-200 kg/ha, Potassium silicate 40-60 kg/ha.
2. In the long term, Si deficiency is prevented by not removing the straw from the field following harvest. Recycle rice straw (5-6% Si) and rice husks (10% Si).
3. Avoid applying excessive amounts of N fertilizer, which increases yield and total uptake of N and Si, but also decreases the Si concentration in straw because of excessive biomass growth.

Silicon Sources

Calcium silicate, potassium silicate, blast furnace slag.

Silicon deficiency

Sodium

Toxicity Symptoms

1. Leaves short, narrow and brittle.
2. Leaves initially show dark green colour, later become bleached and turn yellow to pink and necrotic.

Corrective Measures

Soil application of gypsum at 100-200 kg/ha and leach with water.

Sulphur

Toxicity Symptoms

1. Interveinal chlorosis of emerging leaves.
2. Coarse, sparse, dark brown to black roots.
3. Fresh uprooted rice has poorly developed root systems with many black roots.
4. Increased occurrence of diseases.

Corrective Measures

1. Apply K, P, and Mg fertilizers.
2. Apply Fe (salts, oxides) on low-Fe soils to increase immobilization of H_2S as FeS.
3. Avoid continuous flooding and use intermittent irrigation in soils that contain large concentrations of S.
4. Carry out dry tillage after harvest to increase S and Fe oxidation during the fallow period.

Organic Manures

- Green Manuring.
- Biofertilizer.
- Organic Manuring Practices.

Green Manure

Green Manures	Seed rate (kg/ha)	Green biomass (t/ha)
Sithagathi (Sesbania speciosa)	30 – 40	15-18
Dhaincha (Sesbania aculeata)	50	25
Manila Agathi - Sesbania rostrata	40	20
Sunnhemp (Crotalaria juncea)	25-35	13-15
Wild Indigo (Tephrosia purpurea)	15-20	6-7

Green Manure

Green Leaf Manures

- Glyricidia (Glyricidia maculata Syn. G.sepium).
- Pungam (Derris indica Syn. Pongamia glabra).

Ipomoea Cornea

- Neem (Azadirachta indica).

Sesbania Grandiflora

Glyricidia

Pungam

Neem

Sesbania grandiflora

Advantages of Green Manuring

1. Green manuring builds up soil structure and improves tilth.
2. Improves aeration in rice soils by stimulating activities of surface film of algae and bacteria.
3. Harbour N fixing bacteria, rhizobia in root nodules and fix atmospheric N (60 to 100 kg N/ha).
4. Promotes formation of crumbs in heavy soils leading to aeration and drainage.
5. Vitamin and protein content of rice increased.

Biofertilizer

1. Biofertilizers are ready to use live formulates of such beneficial microorganisms which on application to seed, root or soil mobilize the availability of nutrients by their biological activity in particular, and help to build up the micro-flora and in turn the soil health in general.
2. Bio-fertilizers are eco friendly and are environmentally safe.
3. They form not only part of integrated nutrients but are of low cost.
4. The bio-fertilizers used for rice crop are azolla, Blue green algae, Azotobacter, Azospirillum, Phosphobacteria, Phosphate solubilisers and Mycorhiza.

Azolla

1. Azolla is a fresh water fern.
2. 0.5 -1 t of biomass of azolla/ha is used for paddy in two ways either as green manure before transplanting, or as dual crop in 7 DAT.
3. It releases the nitrogen to the rice crop only after complete decomposition (8-10 days).
4. It also enhances the availability of phosphorus than chemical fertilizer on 40th day besides improving the C:N ratio of soils.

Blue Green Algae

1. These are photosynthetic prokaryotic microorganisms capable of fixing atmospheric nitrogen.
2. It can be applied to rice crop at 10 kg/ha on 10 DAT. 'Algalisation' increases the nitrogen content of both grain and straw besides increasing soil fertility.
3. It can fix 20-30 kg N/ha.

Spirulina-blue-gree-algae

Azotobacter

1. Azotobacter is a free living nitrogen fixing bacteria.
2. It can be applied to rice through seed or seedling or soil.

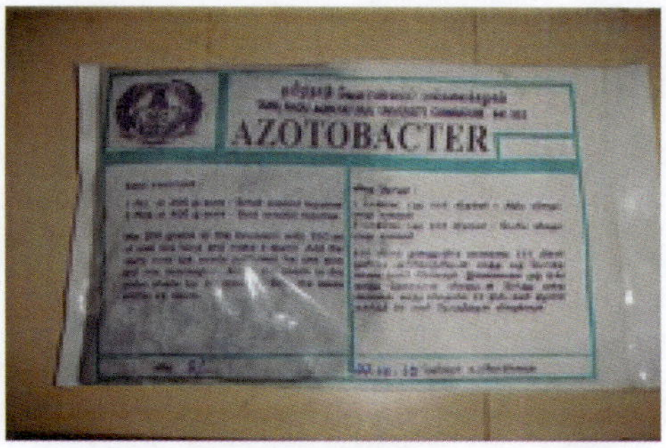

Azotobacter

Azospirillium

Inoculation with Azospirillum promotes early tillering and also the growth of rice and significantly increases filling rate of grain and the grain weight per plant at harvest.

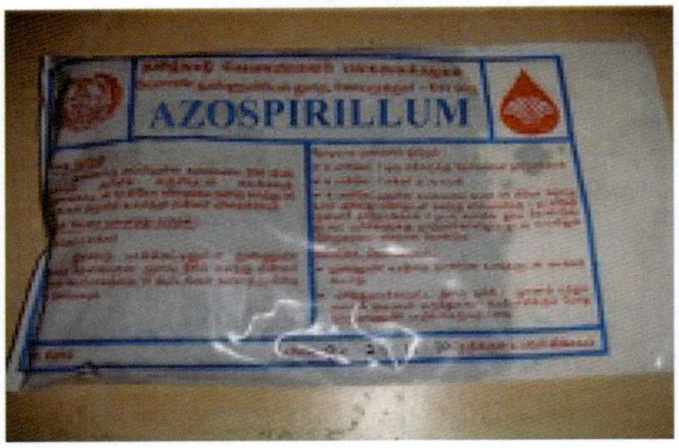

Azospirillium

Phosphobacteria

1. A large proportion of the phosphatic fertilizer applied to soil is fixed by conversion into insoluble forms of phosphorus which is not available to plants.
2. Only 20-25 per cent of phosphorous is available to plants. This insoluble and unavailable phosphorus remains unutilized by plants in soil.
3. The bacteria, which dissolve the undissolved form of phosphorus is called 'Phosphobacteria'.
4. The organic acids produced by phosphobacteria dissolve the insoluble (fixed) form of phosphorous in soil and make it available to plants.

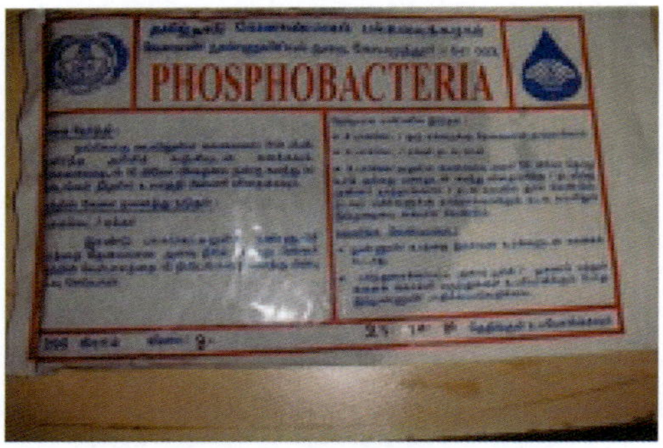

Phosphobacteria

Azospirillium, Phosphobacteria and Azotobacter

Number of packets/hectare is:

Method of application	Number of Packets/ha
Seed treatment	5
Nursery application	10
Seedling dip	5

Main field (Soil application)	10
Total	30

Mycorrhiza

1. It occurs naturally in low land and upland rice. It mobilizes the phosphorus required by rice.
2. It also provides nutrients such as iron, zinc, copper, manganese etc.
3. Endomycorrhiza are obligate symbionts and can be maintained only on live plants inoculated with spores of a species and collecting the pieces of roots with soil.
4. The root biomass heavily infected by a specific mycorrhizal fungus serves as the inoculum for subsequent plots.

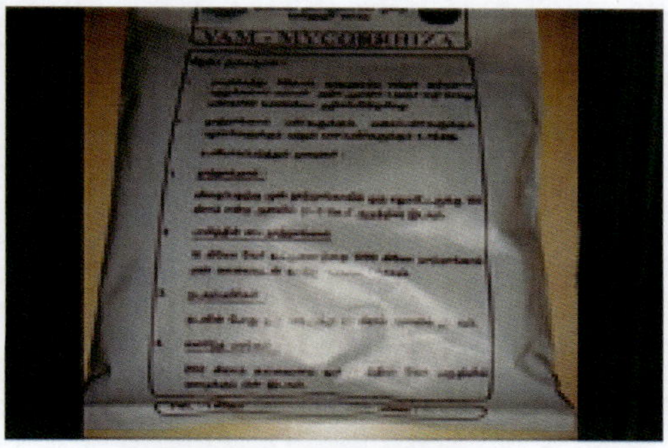

VAM-MYCORRHIZA

Phosphate Solubilisers

1. This type of biofertilizers solubilises phosphates in the soil and render them in available form for low land and upland rice.
2. Bacteria like Bacillus megatherium var phosphaticum, Bacillus polymixa, Pseudomonas fluorscens, Pseudomonas striata fungi like Pencillium digitatum, Aspergillus niger, Aspergillus awamori were found to have a strong phosphate dissolving ability.
3. For semi-dry rice, hardened seeds are treated with Pseudomonas fluorscens 10g/kg of seed.

Phosphate Solubilisers

Methods of Application of Biofertilizer

1. Bacterial biofertilizers are supplied as carrier based inoculants.
2. Peat or lignite is used as carrier material.
3. Carrier based bacterial inoculants are applied by the following methods.
4. Seed treatment.
5. Seedling root dip and
6. Main field application.

Dip the seedings in biofertilizer

Seed Treatment

1. One package of the inoculant is mixed with 200ml of rice kanji to make slurry.
2. The seeds required for an acre are mixed in the slurry so as to have a uniform coating of the inoculant over the seeds and then shade dried for 30 minutes.
3. The shade dried seeds should be sown within 24 hours.
4. One packet of the inoculant (200g) is sufficient to treat 10 kg of seeds.

Seed treatment with biofertilizer

Seedling Root Dip

1. This method is used for transplanted crops.
2. Two packets of the inoculant is mixed in 40 litres of water.
3. The root portion of the seedlings required for an acre is dipped in the mixture for 5 to 10 minutes and then transplanted.

Main Field Preparation

Four packets of the inoculant is mixed with 20 kgs of dried powdered farm yard manure and broadcast in one acre of main field just before transplanting.

Mixing of biofertilizer with FYM for application

Combained Application of Bacterial Biofertilizers

Phosphobacteria can be mixed with Azospirillum. The inoculants should be mixed in equal quantities and applied.

Soil application of biofertilizer

Organic Manuring Practices

1. In nursery, apply 1 tonne of fully decomposed FYM or compost to 20 cents nursery and spread the manure uniformly on dry soil.
2. In main field, apply 12.5 t of FYM or compost or green leaf manure 6.25 t/ha.
3. If green manure is raised 20 kg/ha in situ, incorporate it to a depth of 15 cm using a green manure trampler or tractor.

4. In the place of green manure, press-mud / composted coir-pith can also be used.

Apply 12.5 tones of Farms yard manure

Apply 6.25 tones of green manure

5. Apply organic manure in the form of farmyard manure or compost or green leaf of 5 t/ha and incorporate into the soil while ploughing.
6. Cowpea may be raised as an intercrop in dry seeded low land (semi-dry) rice by sowing 12.5 kg seed/ha along with rice to serve as a source of green manure.
7. When the rice field gets submerged with the onset of southwest monsoon, cowpea at the age of about six weeks and at active vegetative stage decays and gets self-incorporated in the soil adding substantial quantity of green manure.
8. Such a system of concurrent growing of cowpea also reduces weed pressure in semi-dry rice.

Cowpea-source of green manure

Drill Sown Paddy

1. Sow sunhemp green manure seeds 10kg/ha mixed with paddy seeds.
2. Carry out hodta operation (Planking) in standing water after 40 DAS for in situ incorporation of sunhemp in the soil which will decompose easily and early and add to organic matter to the soil maintaining the soil fertility.
3. Ex situ incorporation of green leaf manuring of Eupatorium/parthenium/cassia and other weeds green material 5 t/ha in between the two paddy rows by carrying out hodta operation.

4. Provides only 50% nutrients to maintain good yield.

Transplanted Paddy

Sow green manure seeds of Sesbania rostrata (25kg/ha) along with the application of entire P_2O_5 recommended for paddy, eight weeks before transplanting of paddy, then in situ incorporate the green manure crop by carrying out hodta operation (Planking) seven weeks after sowing.

Transplant paddy seedlings after one week of incorporation along with the application of 50% recommended nitrogen for paddy.

Incorporating sesbania

Sow sunhemp seeds at 10 kg/ha

Fertilizer Requirements

These products are quick acting, even in cool soils and they are inexpensive. They are the most effective means of increasing crop production and supplement nutrient supply in the soil, especially to correct yield-limiting factors.

Transplanted Puddled Low Land Rice

Blanket recommendation (Kg/ha) – 150:50:50.

Stages of application	Split does (Kg/ha)			Straight Fertilizer (Kg/ha)		
	N	P_2O_5	K_2O	Urea	Single super phosphate	Muriate of Potash
Basal	37.5	50	12.5	81.38	312.50	20.75
Active tillering	37.5	-	12.5	81.38	-	20.75
Panicle initiation	37.5	-	12.5	81.38	-	20.75
Heading	37.5	-	12.5	81.38	-	20.75

In nursery, before the last puddling apply basal application of DAP (40 kg) is recommended when the seedlings are to be pulled out in 20-25 days after sowing. For clay soils where root snapping is a problem, 4kg of Gypsum and 1 kg of DAP per cent can be applied at 10 DAS.

Direct wet seeded puddled lowland rice, Dry seeded rainfed un-puddled lowland rice and Rainfed upland rice.

Blanket recommendation (Kg/ha) – 50:25:25.

Stages of application	Split does (Kg/ha)			Straight Fertilizer (Kg/ha)		
	N	P_2O_5	K_2O	Urea	Single super phosphate	Muriate of Potash
Basal	12.5	25	6.5	27.13	156.25	10.38
Active tillering	12.5	-	6.5	27.13	-	10.38
Panicle initiation	12.5	-	6.5	27.13	-	10.38
Heading	12.5	-	6.5	27.13	-	10.38

Semidry System-Dry Seeded Rice in Un-puddled Low Land

Blanket recommendation (Kg/ha) – :75:25:37.5.

Stages of application	Split does (Kg/ha)			Straight Fertilizer (Kg/ha)		
	N	P_2O_5	K_2O	Urea	Single super phosphate	Muriate of Potash
Basal	18.5	25	9.375	40.69	156.25	15.56
Active tillering	18.5	-	9.375	40.69	-	15.56
Panicle initiation	18.5	-	9.375	40.69	-	15.56
Heading	18.5	-	9.375	40.69	-	15.56

Kind of land/region	Variety	Blanket recommendation (Kg/ha)			Straight Fertilizer (Kg/ha)		
		N	P_2O_S	K_2O	Urea	Single Super Phosphate	Muriate of Potash
Uplands (modan)	PTB 28,29,30	40	20	30	86.80	125.00	49.80
	Short duration varieties	60	30	30	130.20	187.50	49.80
Wet lands (All regions)	Short duration varieties	70	35	35	151.90	218.75	58.10
	Medium duration varieties	90	45	45	195.30	281.25	74.70
Kole lands	Short duration verieties	90	35	45	195.30	218.75	74.70
	Medium duration varieties	110	45	45	238.70	281.25	74.70
Kattukampal and Ponnani kole lands	Medium duration varieties	110	45	55	238.70	281.25	91.30

Stages of Application

1. For short duration varieties, N is applied in three equal splits viz., basal, active tillering, and active panicle initiation stages. The recommended P is fully applied as basal. K is applied in two equal splits viz., basal and active panicle initiation stages.

2. For medium and long duration varieties, N are applied in two equal splits viz., basal and active panicle initiation stages. The recommended P is fully applied as basal. K is applied in two equal splits viz., basal and active panicle initiation stages.

3. For direct seeded crop, the basal application should be done one week after sowing.

ZONE	Blanket recommendation (kg/ha)			Straight fertilizers (kg/ha)			Time of application
	N	P_2O_5	K_2O	Urea	Single of potash	Muriat of potash	
1,2,8	100	50	50	217.00	312.50	91.30	N-50% - before transplanting, 25%-30 DAT, 25% panical initation P, K & Zn- Full dose as basal
3	150	75	75	325.50	468.75	124.50	N-50% - before transplanting, 25%-30 DAT, 25% panical initation p,k& Zn- Full dose as basal
9	75	75	90	162.75	468.75	149.40	N-50% - before transplanting, 25%-30 DAT, 55% panical initation p,k& Zn- Full dose as basal
10	75	75	90	162.75	468.75	149.50	N-1/3 - before transplanting, 1/3 -30 1/3-55 days p,k& Zn- Full dose as basal K-1/2 -55 days
8&9 (Drill sown paddy)	100	50	50	217.00	312.50	91.30	N-1/3 – 20 DAS 1/3 40 DAS,panical initation p,k& Zn- Full dose as basal

Zone 1 - North Eastern Transition Zone

Zone 2 - North Eastern Dry Zone

Zone 3 - Northern Dry Zone

Zone 4 - Central Dry Zone

Zone 5 - Eastern Dry Zone

Zone 6 - Southern dry Zone

Zone 7- Southern Transition Zone

Zone 8 - Northern Transition Zone

Zone 9 - Hill Zone

Zone 10 - Coastal Zone

Method of Fertilizer Application

Soil Application

Basal application:

1. Apply 25 % recommended dose of N and K as basal.
2. P may be applied fully as basal and incorporated.
3. Apply 25 kg of zinc sulphate mixed with 50 kg dry sand just before transplanting.
4. Apply 500 kg of gypsum/ha (as source of Ca and S nutrients) at last ploughing.

Top Dressing

Apply 25 % recommended dose of N and K each as top dressing at active tillering, panicle initiation and heading stages.

Soil application

Foliar Nutrition in Paddy

1. In recent year's soluble fertilizers otherwise known as foliar fertilizers like Polyfeed and Multi 'K' were introduced in rice growing states.
2. Polyfeed contains 19: 19: 19 NPK with 6 micro-nutrients like iron, manganese, boron, zinc, copper and molybdenum, while multi K contains 13: 0: 46 NPK.
3. These fertilizers provide nutrients to the plant by foliar application as these fertilizers are completely soluble in water.
4. These fertilizers have no other impurities like sodium and chloride and they are 100 % nutrients and these nutrients are easily absorbed through the leaves.
5. In certain occasions like prolonged drought, there is no scope to apply fertilizers to the soil for want of moisture. Like wise in flooded conditions due to continuous rains fertilizers could not be applied to the soil. In such special circumstances these soluble fertilizers are must to protect the crop against hunger and this forms a compulsory act of crisis management.
6. And the foliar application of Speciality fertilizers plays an important role in supplying the nutrients at critical stages of flowering and grain formation.
7. Under special conditions of drought and waterlogging, apply N as foliar spray.
8. Urea may be applied as a low volume spray at 15% concentration using power sprayer or at 5% concentration using a high volume sprayer, the quantity applied in one application being limited to 15 kg/ha.
9. Foliar spray of Urea (10 g/lit) + DAP (20 g/lit) + KCl (10 g/lit) at PI and 10 days later for all varieties.
10. If deficiency symptom appears, foliar application of 0.5% Zinc sulphate + 1.0% urea can be given at 15 days interval until the Zn deficiency symptoms disappear.

Growth Regulators

1. Foliar spray of Brassinosteriods 0.3 ppm at Panicle Initiation and Flowering stages increased the grain yield.

2. For increasing the rooting under broadcast method of planting, soaking roots in 25 ppm Induction of better rooting for early establishment in rice, root dipping for 16 hours in thiamin solution.

Spraying of growth regulator

Nutrient Management Planning

Nutrient management traditionally has been concerned with optimizing the economic returns from nutrients used to produce a crop. More recently, nutrient management also has begun to address ways to minimize the negative impact of nutrients on the environment.

The environmental problems associated with nutrients most often are caused by animal manure. Applying manure in excess of crop needs or at the wrong time, or handling it improperly may release nutrients into the air or water, where they no longer contribute to crop production and may act as pollutants. The leaching of nitrogen through the soil can raise groundwater nitrate levels. In addition, runoff and erosion may increase nitrogen and phosphorus levels in surface waters, which can lead to eutrophication and related problems such as algae scums, odors, and loss of fish populations. Good nutrient management planning can help to avoid some of these problems.

Nutrient management plans are not new. All farmers have a plan for using the manure produced by their animals. In many cases, however, this plan is very informal and ad- dresses only manure disposal and possibly the crop response to the manure nutrients; environmental concerns usually are not addressed. Changing regulations now require farmers to implement more formal nutrient management plans that address environmental issues.

A basic nutrient management plan includes the following:

1. An inventory of nutrient sources on the farm, including manure and crop residues.
2. Nutrient analyses of each of these sources. The compositions of many of these sources may vary considerably, so "book values" may be inadequate for planning purposes.

3. Handling and storage procedures for minimizing the potential for nutrient loss around the barnyard.
4. Lists of crops and crop rotations by field, or crop groups and expected yields.
5. Soil tests to determine the nutrient needs of the crops to be grown.
6. Prioritization of the fields based on maximizing the economic benefit and minimizing the environmental impact of the manure nutrients.
7. Procedures for when and how to apply the manure to maximize the economic benefit and minimize the environmental impact of the nutrients.
8. A practical manure spreading rate for each field that does not apply the nutrient of greatest environmental concern in excess of the crop needs.
9. Best management practices that minimize the potential for nutrient loss from the fields.
10. Nutrient balances for each field to indicate any additional nutrient needs for the crop and any excesses that might be of concern.
11. Plans for dealing with any excess manure produced on the farm.

While all manure nutrient management plans should contain these basic plan elements, the actual plans themselves will take on many forms.

Developing a Nutrient Management Plan

A nutrient management plan can be written with paper and a pencil or developed with computer software. The key concept is that the plan allocates the available manure nutrients in a way that maximizes the economic benefit of the nutrients while minimizing their environmental impact. The basic steps in developing a nutrient management plan are as follows:

The first step in developing a plan for manure management is to collect information about the manure production, nutrient content, and application systems for the farm.

Many factors affect the amount of manure produced and the nutrients it contains. These include the type of animal and its age, ration, and feed consumption; the animal's management, including the bedding used and the length of time the animal is confined as the manure is collected; and the handling of the manure before and during field application, including potential drying and dilution by waste water or precipitation.

Manure production can best be determined by measuring the amount of manure contained in a manure storage. On farms that do not have a storage or when it is impossible to determine the amount in a storage, manure production can be estimated from animal numbers, animal weights, and confinement times.

Manure nutrient content should be determined by a manure analysis. "Book" values for the nutrient content of a given type of manure are averages taken from many farms. Because of farm-to-farm variability, these averages are of little value in decision making on an individual farm. Manure sample analysis results in Pennsylvania have indicated a very wide range in nutrient content for seemingly similar samples of manure taken from different farms, and we have observed as much as 100 percent error between book values and actual manure nutrient content values.

The second step is to determine the nutrient requirements for the crops on the farm. The nutrient requirements preferably for individual fields or groups of fields with the same crop and very similar rotation and nutrient application histories must be determined. Soil tests are the best source of information about crop nutrient requirements. Soils contain a wide range of available nutrient levels, and different crops have very different nutrient requirements. The expected yield of the crop also will affect the amount of nutrient required. For example, a corn crop requires a large amount of nitrogen and a smaller amount of phosphate and potash. An alfalfa crop, however, requires no nitrogen, some phosphate, and a large amount of potash.

In general, forage crops such as hay or corn silage will use more nutrients than grain crops. A rotation of grain and forage crops will require nutrients very different from those needed by either of the individual crops grown in a field in a given year. These facts become very important when manure is used to meet crop nutrient needs. Tables of crop nutrient requirements are available. All sources of nutrients must be considered in developing a manure application plan. Other sources of nutrients must be subtracted from the crop requirements to determine a net nutrient requirement before a manure rate is determined. These might include: fertilizer applied regardless of manure applications, such as starter fertilizer; residual nitrogen from a previous legume crops; and residual nitrogen from past manure applications.

The third step is to prioritize the farm fields on the basis of their suitability for manure application. Using soil tests, the fields' cropping and manure application history, the planned crop rotation, and other characteristics such as field location, slope, and soil characteristics, rank the fields for manure applications from highest to lowest.

The fourth step is to determine the availability of manure nutrients. The actual fertilizer value of the manure will depend on how the manure is handled and used. The nutrients in manure are not as readily available as fertilizer nutrients are, and this must be taken into account when manure application rates are determined. Manure nitrogen is especially susceptible to losses and is very sensitive to management practices such as soil incorporation following application. For example, if dairy manure is incorporated on the same day that it is spread, about half of its N will be as readily available as commercial fertilizer N would be. If this manure is left on the surface with no incorporation for a week or more, however, only 20 percent of the N will be available for crop uptake in the current year. Tables and calculation worksheets are available to help you estimate the availability of the manure nutrients in your management system.

The fifth step is to determine the appropriate manure application rate and allocate the available manure to the fields in the established priority order. The rate calculations are based on meeting the net nutrient requirement for the nutrient of principal concern (usually N or P). If the rate is based on N, the manure analysis must be adjusted for N availability. The best approach to determining an environmentally sound manure application rate is to use the rate that does not apply any nutrient in excess of the soil test's recommendations. In our experience, P is usually the limiting nutrient when this approach is followed.

For fields where there is a crop rotation, base the calculations for this limit on the nutrient needs of the entire rotation rather than on the current crop only. In Pennsylvania, we typically base the manure application rate on N and try to minimize excess P and K applications based on the rotation requirements. In practice, many farmers still apply excess P and K. Since the major loss pathways for P and K are field runoff and erosion, good soil conservation practices are critical in minimizing the environmental effect of excess P and K applications in manure.

The sixth step is to adjust the calculated manure rates for practicality. This usually results in the grouping of the fields into one or a few standard manure application rates that the farmer is able and willing to apply. Manure spreader calibration information is a good guide for determining these practical rates. Using these standard rates, the available manure is allocated to the fields in priority order until all of the manure is allocated or all of the fields have been fertilized. If manure remains, a plan is developed to remove it from the farm, or the cropping program is adjusted to use up small excess amounts.

The seventh step is to compare the nutrients supplied in the manure to the needs of the crop to determine if additional nutrients are required or if a serious excess of any nutrient is being applied. Deficiencies can be handled by applying supplemental fertilizer nutrients or manure from other sources. Serious excesses must be evaluated in light of the nutrient needs of the crop rotation and the potential impact area of the nutrient loss, and may require a change in the manure application plan. The Phosphorus Index (PI) is a good tool to evaluate the risk of excess P loss to the environment.

Finally, the plan must include appropriate best management practices (BMPs) to protect water quality from nutrient contamination. These BMPs include practices such as manure handling and storage, barnyards, animal concentration areas, and stormwater runoff from field application areas. A current, implemented farm conservation plan is an important compliment to a nutrient management plan.

Table: Factors involved in field prioritization for manure applications.

Nutrient Considerations		
Nutrient	Prioritization for Manure	
	High	Low
Crop N needs	N-requiring crops	Crops not requiring N
N requirement	Highest N requirement	Lowest N requirement
Residual N	Lowest residual N	Highest residual N
P soil test level	Lowest P level	Highest P level
K soil test level	Lowest K level	Highest K level
Environmental Considerations: Proximity to water bodies, sinkholes, flood plains, and drinking water sources/wellhead protection areas.		
Soil Limitations: Leaching, erodibility, and runoff potential.		
Vegetative Cover: Presence of a growing crop, crop residue, or cover crop.		
Other Considerations:Distance from source to fields, neighbors, etc.		

Implementation of the Nutrient Management Plan

- Manure always should be spread on fields uniformly and at the planned rates.
- Manure spreader calibration is the key for implementing the nutrient management plan. Procedures for calibrating different types of manure spreaders are available.
- Crop nutrient uptake is most efficient if manure is spread as close as possible to the time when plants will use the nutrients. The longer the manure is in the soil before the crops take up the nutrients, the more susceptible those nutrients will be to loss.
- The season in which manure is applied will affect the nutrient availability for crops. The best time is usually in the spring, just before the crop will use the nutrients. Summer also is good if application to growing crops is possible. Fall and winter are the riskiest times to spread because there is little crop growth to use the nutrients until the following growing season. A higher potential for nutrient loss also exists during these seasons because of frozen soils and high precipitation. If manure must be applied in the fall and winter, the use of a cover crop is highly recommended.
- Incorporation of manure into the soil as soon as possible after it is spread results in the greatest nutrient availability to crops and reduces nutrient losses, pollution, and odor. One-half inch of soaking rainfall without significant runoff is comparable to incorporation of surface-applied manure.
- Surface-applied manure may be lost through surface runoff. On soils with a high erosion potential, conservation practices such as stripcropping, contour farming, crop residue management, cover crops, cropland terraces, diversions, and grassed waterways can effectively reduce the loss of soil and manure nutrients.
- Be especially careful with manure management in environmentally sensitive areas such as near streams, water bodies, or sinkholes; in areas where the potential for concentrated water flow exists, particularly in the winter; and near wells and springs. Nutrient application buffer zones may be needed near these sensitive areas.
- Since the amount of manure N that becomes available in any given year is very difficult to predict, the Pre-Sidedress Soil Nitrate Test (PSNT) or the early-season chlorophyll meter test are useful tools that can help determine if the manure N is adequate to meet the needs of a corn crop.
- Manure spreaders have been implicated as a major cause of soil compaction on many farms. Avoid compaction by staying off wet soils with heavy equipment.
- Keep good records of manure application. This will not only document your good management, but also provide valuable information you can use to make future plan adjustments.

Integrated Nutrient Management

Integrated nutrient management is the combined application of chemical fertilizers along with organic resource materials like, organic manures, green manures, bi-fertilizers and other organic decomposable materials for crop production.

The basic concepts of IPNS is the maintenance or adjustment of soil fertility and supply of plant nutrients to an optimum level for sustaining desired crop productivity through optimization of benefits from all possible sources of plant nutrients in an integrated manner.

IPNS is ecologically, socially and economically viable and environment friendly which can be practiced by farmers to derive higher productivity with simultaneously maintaining soil fertility. Integrated nutrient management encourages the use of on-farm organics, thus it saves on the cost of fertilizers for crop production.

The basic concept of integrated nutrient management (INM) or integrated plant nutrition management (IPNM) is the adjustment of plant nutrient supply to an optimum level for sustaining the desired crop productivity.

It involves proper combination of chemical fertilizers, organic manure, crop, residues, N_2~fixing crops (like pulses such as rice bean, Black gram, other pulses and oilseeds such as soybean and bio-fertilizers suitable to the system of land use and ecological, social and economic conditions.

The cropping system rather than an individual crop, and farming system rather than an individual field, is the focus of attention in this approach for development INM practices for various categories. The basic concept of INM is the maintenance of soil fertility, sustainable agricultural productivity and improving profitability through judicious and efficient use of fertilizers as mentioned.

Concepts of INM

INM use Five Major Sub-concepts

1. Plant nutrients stored in the soil.
2. Plant nutrients, those present in the crop residues, organic manure and domestic wastes.
3. Plant nutrients purchased or obtained from outside the farm.
4. Plant nutrient looses e.g. those removed from the field in crop harvest and lost from the soil through volatilization (ammonia and nitrogen oxide gases and leaching (nitrate, sulphate etc).
5. Plant nutrient outputs e.g. nutrient uptake by the crops at harvest time.

The integrated plant nutrition management on integrated nutrient management thus firstly operates at plot level, optimizing the utilization of plant nutrients from diverse sources which are locally available, in order to improve the agronomic efficiency of such nutrient and at the same time reducing the losses of nutrients. Organic recycling is especially promoted through INM for facilitating nutrient recycling to improve soil fertility and productivity as low cost.

Goals of INM

To Maintain Soil Poductivity

To ensure productive and sustainable agriculture. To reduce expenditure on costs of purchased inputs by using farm manure and crop residue, etc. To utilize the potential benefits of green manures, leguminous crops and bio-fertilizers. To prevent degradation of the environment. To meet the social and economic aspiration of the farmers without harming the natural resource base of agricultural production.

Implementation of INM Activities

Different stages of implementation of INM are as follows:

1. Diagnosis phase: collection of background information.
2. Analysis of constraints.
3. Preparing potentiality and feasibility summary.
4. On-farm demonstrations.
5. Evaluation of INM activities.

Components of INM

Components of INM and their use: Major components of integrated nutrient management are:

1. Integration of soil fertility restoring crops like green manures, legumes etc.
2. Recycling of crop residues.
3. Use of organic manures like FYM, compost, vermicompost, biogas, slurry, poultry manure, bio-compost, press mud cakes, phosphocompost.
4. Utilization of Bio fertilizers.
5. Efficient genotypes and lastly.
6. Balanced use of fertilizer nutrients as per the requirement and target yields.

Diagnostic Phase of INM

In the first stage of diagnostic phase information with regard to the following is collected and analysed:

1. Farming/cropping systems.
2. Crop varieties grown.
3. Awareness about soil fertility problems.
4. Use of chemical fertilizers, lime/dolomite and other agro-chemicals.
5. Use of organic manure.
6. Availability of fertilizer and other inputs.
7. Irrigation sources and practices.
8. Soil testing service facility.

9. Constraints in the adoption of INM technologies.
10. Consideration of market opportunities.

The INM technologies must be compatible with the local farming system if they are to find acceptance and adoption. Therefore, attention must be paid to examine the interaction among different components of INM and the management of crops and animals that form the farming system.

Some of the agronomic parameters which need attention are cropping pattern, intercropping practices, biological condition of the field (weeds, diseases, and insects), soil conditions, irrigation facilities and climatic conditions.

Common Constraints of INM

Common constraints encountered by the farmers in adoption of INM technology are as follows:

1. Non-availability of FYM.
2. Difficulties in growing green manure crops.
3. Non-availability of bio-fertilizers.
4. Non-availability of soil testing facilities.
5. High cost of chemical fertilizers.
6. Non-availability of water.
7. Lack of knowledge and poor advisory services.
8. Non-availability of improved seeds.

Integrated Plant Nutrient Management at Farm Level

It is only after they have made improvements in the biological, physical and hydrological properties of their soils that farmers can expect to get the full benefits from the supply of additional plant nutrients, in the form of inorganic fertilizer, to their crops. At the farm field level IPNM therefore calls for an integrated and synergistic approach which involves:

1. Matching the land use requirements of individual agricultural enterprises with the land qualities present in the areas where they are undertaken - i.e. the biological, chemical and physical properties of the soil, and the local climatic conditions (temperature, rainfall etc).
2. Seeking to improve yields by identifying and overcoming the most limiting factors in order of their diminishing influence on yield.
3. Better plant management, especially: (i) improved crop establishment at the beginning of the rains, so as to increase protective ground cover thereby reducing splash erosion, enhancing infiltration and biological activity; and (ii) timely weeding to reduce crop yield losses from competition for nutrients and soil moisture.

4. Combinations of complementary crop, livestock and land husbandry practices which maximize additions of organic materials and recycle farm wastes, so as to maintain and enhance soil organic matter levels (ideally at levels of at least 50-75% of those under natural vegetation).

5. Land management practices that ensure soil moisture conditions are favourable for the proposed land use (e.g. moisture harvesting/conservation in low rainfall areas, drainage in high rainfall areas).

6. The replenishment of soil nutrients lost by leaching and removed in harvested products through an integrated plant nutrition management approach that optimizes the benefits from all possible on- and off-farm sources of plant nutrients (e.g. organic manures, crop residues, rhizobial N-fixation, P and other nutrient uptake through root mycorrizhal fungi infestation, transfer of nutrients released by weathering in the deeper soil layers to the surface via tree roots and leaf litter, rock phosphate, inorganic fertilizer etc).

7. Combinations of crop, livestock and land husbandry practices that reduce rainfall impact, improve surface infiltration, and reduce the velocity of surface runoff thereby ensuring any soil loss is below the 'tolerable' level for the soil type.

8. Conservation tillage, crop rotation, agroforestry and restorative fallow practices that maintain and enhance the soils physical properties through maintaining an open topsoil structure, and breaking any subsoil compacted layer (hoe/plough pan) thereby encouraging root development and rainfall infiltration (e.g. use of ox drawn chisel ploughs, double dug beds, pasture leys, interplanting of deep rooted perennial crops/trees and shrubs).

Good Agricultural Practices for Nutrient Management

Integrated plant nutrient management also contributes to pest management stressed crops are more susceptible to disease and to the effects of pest attacks. Crops growing in poorly structured soil, under low or unbalanced nutrient conditions or with inadequate water supply or retention will be stressed. Responding to disease or pest attacks by applying pesticides is a costly symptomatic approach to a syndrome which is better addressed by improving the ecological conditions and systems within which the crops are cultivated. In addition, agricultural products with less pesticides residues are less risky to consume, and healthy plants with a properly balanced nutrient supply provide better quality feed and food, improving animal and human health.

Production efficiencies are gained through the integrated nutrient management practices promoting combined use of mineral, organic and biological resources in a reasoned way to balance efficient use of limited/finite resources and ensure ecosystem sustainability against nutrient mining and degradation of soil and water resources. For example, efficient fertiliser use requires that correct quantities be applied (overuse of Nitrogen (N) fertilizer risks disrupting the natural N-cycle), and that the application method minimizes losses to air and water. Options exist for incorporating fertilizer into the soil directly (rather than broadcasting). Equally, plant nutrient status during the growing season can be better monitored using leaf-colour charts, and adaptively managing fertilizer application accordingly.

Nutrients are in many ways linked to other practices that are related to sustainable crop production intensification. Nutrient cycling is the main focus of Conservation Agriculture, in which minimum soil disturbance, intercropping, crop rotations and a permanent soil cover minimize the need for chemical fertilizers. Crops with sufficient nutrients are also less susceptible to pests, thus contributing to Integrated Pest Management. High application of fertilizers can increase the occurrence of noxious weeds that compete with crops for nutrients. This means that Integrated Weed Management is promoted by improving the timing, dosing and application method of nutrients and thus minimizing the potential impact on weed growth. A better application of nutrients also reduces runoff, and by this the overall agricultural biodiversity and the pollination services.

Nutrient Management in Crop-livestock Systems

The management of nutrients is one of the most important, but also complicated issues in crop-livestock systems. Nutrients can come from manure, legumes, and external inputs. A closed cycle – without external inputs - would be ideal from an environmental point of view, but might not always be feasible due to soil constraints, difficulties for growing legumes or other nitrogen fixing crops, or lack of options for spreading the manure around the farm (distance between plots or size of farm). Recycling can be done to effectively use both on-farm and off-farm wastage. For crop and fodder production a combination of mineral and organic fertilizers can be applied from on-farm and off-farm sources. The crop rotations in the crop-livestock systems would preferably also include crops that are beneficial from a nutrient point of view (e.g. legumes). A balanced nutrient level in the soil will contribute to sustainable crop (and livestock) production intensification – this means that a focus should be on farm level, and not on the fertilization of single crops. The focus of farmers should be on the total farm area and on the long-term effect of fertilization. This includes making a shift from static nutrient balances to nutrient flows in cycles.

Since the demand for livestock will inevitably continue to grow over the next decades, it will be very important to deal with this increased production in a sustainable manner. Integrating crop with livestock provides benefits to the environment by created a 'closed' system for nutrients and to animal welfare by providing sufficient space for the animals to graze. Whereas the increased production in the past has mainly come from increasing the area for livestock and shifting towards industrial systems, both are facing increasing limitations; the area for expansion is limited and industrial systems are seeing increasing opposition due to the accompanying ethical dilemma caused by the perceived lower animal health and welfare in these systems. Managing crop-livestock systems in a way that they will be environmentally sustainable and have an increased stocking density will be one of the main challenges in agriculture development in the coming decades. This will include improving the livestock breeds and management of feed, land, and water. High pressures of livestock on pasture areas (which are often degraded crop-land), can cause further degradation of land and water resources and create conflicts with other sectors.

Soil Health

Soil health is a condition, or status, of the soil at a certain place and in a specific environment as compared to a certain reference or benchmark condition. However, the concept of soil health can vary in use based on the priorities placed on different soil functions. Therefore, the concept of soil health should be understood within the context and intention of the users of the soil health term, their goal and the boundaries in which they are working.

In general, soil health, as a measure of soil functions, can be defined as the optimum status of the soil's biological, physical and chemical functions. This means healthy soils can sustain plant and animal productivity and soil biodiversity maintain or enhance water and air quality, and support human health and wildlife habitat. The underlining principle for a healthy soil is not just a medium to grow plants, but rather a living, dynamic and changing environment that is influenced by what we do and the practices we adopt as human activities. If we agree that the soil is a dynamic and living organism, then we need to apply the human health principles as its metaphor. We, as humans, strive for healthy bodies to perform our tasks most efficiently, and so do soils.

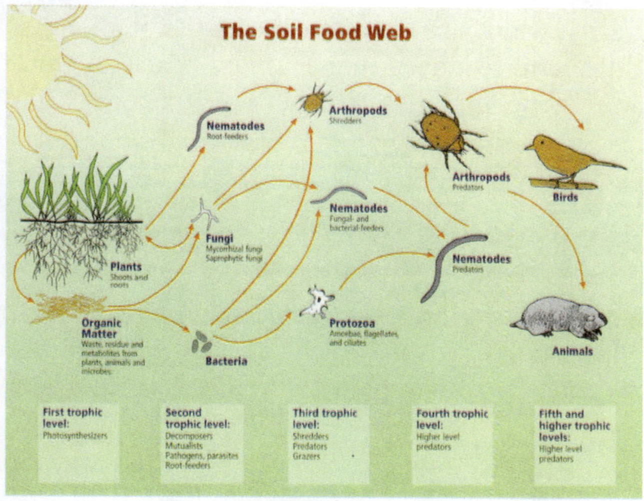

Soil food web

To understand soil health and how different human activities (i.e., cultivation, land use, etc.) impact it, we need to understand the soil. The soil is a heterogeneous natural body, which basically consists of solid particles (mineral particles), organic matter, water and air. The soil also contains micro and macro organisms that support plant, animal and human life. Its functions as a heterogeneous natural body consist of providing support as the medium for plant growth, the storage and supply of water and nutrients, the purification of pollutants, and wildlife habitat. Organic matter is one of the smallest components of the soil system, but plays an essential role in maintaining soil health/functions. Soil organic matter is derived from living organisms, such as plants and animals, and their byproducts in the soil environment. When organic matter breaks down, it is transformed into different pools as sources of plant nutrients at various degrees of availability and eventually forms the final product called humus. This product becomes the central building block of healthy soil. Therefore, the maintenance of soil organic matter is critical to the health and productivity of the soil; providing a stable soil physical structure for water storage, nutrient exchange with plant roots, aeration and a healthy microbial community will enhance soil health for healthy plant growth.

In agricultural systems, such as row cropping systems, significant stress is exerted on soil functions through management practices such as soil tillage, chemical application and continuous mono-cropping systems. However, management practices, such as soil conservation systems, including no-tillage and extended crop rotations can mitigate negative effects on soil health/functions. A no-till system can restore soil health over time by improving, for example, soil infiltration, organic matter, water storage, soil structure, etc., which are indicators of soil health. The extended crop rotations that include small grains, legumes and cover crops will increase soil biodiversity and protect the soil surface physically during the off season and provide organic carbon input. The introduction of perennials on marginal land can increase wildlife habitat and improve the biological and physical components of soil health. These practices are measures to build healthy soil, which can improve both productivity and the environment.

When the soil is managed as a natural heterogeneous ecosystem by diversifying the input of plants, animals and wildlife, we incorporate the natural order of soil generation principles that led to the forming of the soil. The challenges we face in many parts of the world nowadays are the outcomes of human activities that affect soils by accelerating soil degradation and soil erosion and decreasing water quality and quantity, which are future food security challenges. These challenges can be mitigated through thoughtful and balanced management practices that enhance soil health sustainably as the first step in establishing potential solutions for these problems.

Soil Health Assessment and Management

Soil health is an assessment of how well soil performs all of its functions now and how those functions are being preserved for future use. Soil health cannot be determined by measuring only crop yield, water quality, or any other single outcome. Soil health cannot be measured directly, so we evaluate indicators.

Conservationists learn how to assess soil quality

Indicators are measurable properties of soil or plants that provide clues about how well the soil can function. Indicators can be physical, chemical, and biological properties, processes, or characteristics of soils. They can also be morphological or visual features of plants.

Useful indicators:

- Are easy to measure,
- Measure changes in soil functions,
- Encompass chemical, biological, and physical properties,
- Are accessible to many users and applicable to field conditions, and
- Are sensitive to variations in climate and management.

Indicators can be assessed by qualitative or quantitative techniques. After measurements are collected, they can be evaluated by looking for patterns and comparing results to measurements taken at a different time or field.

Indicator examples and relationship to soil health:

- Soil Organic Matter: Nutrient retention; soil fertility; soil structure; soil stability; and soil erosion.
- Physical: Bulk density, infiltration, soil structure and macropores, soil depth, and water holding capacity, retention and transport of water and nutrients; habitat for soil microbes; estimate of crop productivity potential; compaction, plow pan, water movement; porosity; and workability.
- Chemical: Electrical conductivity, reactive carbon, soil nitrate, soil pH, and extractable phosphorus and potassium, biological and chemical activity thresholds; plant and microbial activity thresholds; and plant available nutrients and potential for N and P loss.
- Biological: Earthworms, microbial biomass C and N, particulate organic matter, potentially mineralizable N, soil enzymes, soil respiration, total organic carbon, microbial catalytic potential and repository for C and N; soil productivity and N supplying potential; and microbial activity measure.

Soil Health Management

Soil works for you if you work for the soil by using management practices that improve soil health and increase productivity and profitability immediately and into the future. A fully functioning soil produces the maximum amount of products at the least cost. Maximizing soil health is essential to maximizing profitability. Soil will not work for you if you abuse it.

Managing for soil health (improved soil function) is mostly a matter of maintaining suitable habitat for the myriad of creatures that comprise the soil food web. This can be accomplished by disturbing the soil as little as possible, growing as many different species of plants as practical, keeping living plants in the soil as often as possible, and keeping the soil covered all the time.

Crop residue management improves soil quality

Manage More by Disturbing Soil Less

Soil disturbance can be the result of physical, chemical or biological activities. Physical soil disturbance, such as tillage, results in bare and compacted soil that is destructive and disruptive to soil microbes, and it creates a hostile environment for them to live. Misapplication of farm inputs can disrupt the symbiotic relationships between fungi, other microorganisms, and plant roots. Overgrazing, a form of biological disturbance, reduces root mass, increases runoff, and increases soil temperature. All forms of soil disturbance diminish habitat for soil microbes and result in a diminished soil food web.

Diversify Soil Biota with Plant Diversity

Plants use sunlight to convert carbon dioxide and water into carbohydrates that serve as the building blocks for roots, stems, leaves, and seeds. They also interact with specific soil microbes by releasing carbohydrates (sugars) through their roots into the soil to feed the microbes in exchange for nutrients and water. A diversity of plant carbohydrates is required to support the diversity of soil microorganisms in the soil. In order to achieve a high level of diversity, different plants must be grown. The key to improving soil health is ensuring that food and energy chains and webs consist of several types of plants or animals, not just one or two.

Biodiversity is ultimately the key to the success of any agricultural system. Lack of biodiversity severely limits the potential of any cropping system and increases disease and pest problems. A diverse and fully functioning soil food web provides for nutrient, energy, and water cycling that allows a soil to express its full potential. Increasing the diversity of a crop rotation and cover crops increases soil health and soil function, reduces input costs, and increases profitability.

Keep a Living Root Growing throughout the Year

Living plants maintain a rhizosphere, an area of concentrated microbial activity close to the root. The rhizosphere is the most active part of the soil ecosystem because it is where the most readily available food is, and where peak nutrient and water cycling occurs. Microbial food is exuded by plant roots to attract and feed microbes that provide nutrients (and other compounds) to the plant at the root-soil interface where the plants can take them up. Since living roots provide the easiest source of food for soil microbes, growing long-season crops or a cover crop following a short-season crop, feeds the foundation species of the soil food web as much as possible during the growing season.

Healthy soil is dependent upon how well the soil food web is fed. Providing plenty of easily accessible food to soil microbes helps them cycle nutrients that plants need to grow. Sugars from living plant roots, recently dead plant roots, crop residues, and soil organic matter all feed the many and varied members of the soil food web.

Keep the Soil Covered as much as Possible

Soil cover conserves moisture, reduces temperature, intercepts raindrops (to reduce their destructive impact), suppresses weed growth, and provides habitat for members of the soil food web that spend at least some of their time above ground. This is true regardless of land use (cropland, hayland, pasture, or range). Keeping the soil covered while allowing crop residues to decompose (so their nutrients can be cycled back into the soil) can be a bit of a balancing act. Producers must give careful consideration to their crop rotation (including any cover crops) and residue management if they are to keep the soil covered and fed at the same time.

References

- Nutrient-management, practices: soilquality.org, Retrieved 11 June, 2019
- Nutrientmanagement, paddy, expert-system: agritech.tnau.ac.in, Retrieved 13 July, 2019
- Nutrient-management-planning, nutrient-management-general, educational, nutrient-management, programs: psu.edu, Retrieved 15 August, 2019
- Integrated-nutrient-management-inm-meaning-concept-and-goals, soil-fertility: soilmanagementindia.com, Retrieved 17 January, 2019
- Integrated-plant-nutrient-management, managing-ecosystems, crops agriculture: fao.org, Retrieved 18 February, 2019
- What-soil-health-and-how-can-we-improve-it: extension.iastate.edu, Retrieved 23 March, 2019
- Health, soils: nrcs.usda.gov, Retrieved 25 April, 2019
- Health, soils: nrcs.usda.gov, Retrieved 24 May, 2019

Permissions

All chapters in this book are published with permission under the Creative Commons Attribution Share Alike License or equivalent. Every chapter published in this book has been scrutinized by our experts. Their significance has been extensively debated. The topics covered herein carry significant information for a comprehensive understanding. They may even be implemented as practical applications or may be referred to as a beginning point for further studies.

We would like to thank the editorial team for lending their expertise to make the book truly unique. They have played a crucial role in the development of this book. Without their invaluable contributions this book wouldn't have been possible. They have made vital efforts to compile up to date information on the varied aspects of this subject to make this book a valuable addition to the collection of many professionals and students.

This book was conceptualized with the vision of imparting up-to-date and integrated information in this field. To ensure the same, a matchless editorial board was set up. Every individual on the board went through rigorous rounds of assessment to prove their worth. After which they invested a large part of their time researching and compiling the most relevant data for our readers.

The editorial board has been involved in producing this book since its inception. They have spent rigorous hours researching and exploring the diverse topics which have resulted in the successful publishing of this book. They have passed on their knowledge of decades through this book. To expedite this challenging task, the publisher supported the team at every step. A small team of assistant editors was also appointed to further simplify the editing procedure and attain best results for the readers.

Apart from the editorial board, the designing team has also invested a significant amount of their time in understanding the subject and creating the most relevant covers. They scrutinized every image to scout for the most suitable representation of the subject and create an appropriate cover for the book.

The publishing team has been an ardent support to the editorial, designing and production team. Their endless efforts to recruit the best for this project, has resulted in the accomplishment of this book. They are a veteran in the field of academics and their pool of knowledge is as vast as their experience in printing. Their expertise and guidance has proved useful at every step. Their uncompromising quality standards have made this book an exceptional effort. Their encouragement from time to time has been an inspiration for everyone.

The publisher and the editorial board hope that this book will prove to be a valuable piece of knowledge for students, practitioners and scholars across the globe.

Index